Laboratory Manual
to accompany

REFRIGERATION AND AIR CONDITIONING
An Introduction to HVAC/R
Fourth Edition

Prepared by:

David Fearnow
Grand Rapids Community College

PEARSON

Prentice
Hall

Upper Saddle River, New Jersey
Columbus, Ohio

10 9 8 7 6 5 4 3 2 1

PREFACE

Laboratory Manual to accompany Refrigeration and Air Conditioning: An Introduction to HVAC/R, Fourth Edition is filled with 92 all-new laboratory worksheets and was written for students in post-secondary training to become HVAC/R technicians. Unlike other manuals that may offer make-work labs with little practical value, this laboratory manual will give the student hands-on experience directly related to work in the field. The student who completes a course using this laboratory manual will have experience closer to an on-the-job apprenticeship than a typical school laboratory.

To the teachers who are using this laboratory manual for the first time, you will find that these laboratory worksheets will be reminiscent of jobs done in the field. Many of you have extensive work experience and are well qualified for work in the field. You will find the 92 laboratory worksheets an essential part of your laboratory course. These labs are divided into seven sections: Air Conditioning, Controls, Gas Furnace, Oil Furnace, Electric Furnace, Refrigeration, and Gas Heating Boiler Laboratories. The two longest sections, Refrigeration and Air Conditioning, each have more than 20 labs that offer direct practical value to the student.

The laboratory worksheets in this manual are innovative and give valuable information found nowhere else in this form. For example, Basic Spring AC Startup is a summation of several factory initial startup sheets, but it goes beyond initial startup. You probably would not be doing "oil all motors as required" on a new piece of equipment. As an instructor I feel that one good lab is worth more than 20 make-work type labs. I have attempted to bridge the gap between actual job experience and generic labs that can be performed on an assortment of typical HVAC/R training equipment.

Some of the labs are basic and entry level, some are more intermediate, and some are rather in-depth, requiring skill and understanding. For example, Laboratory Worksheet GF-2 (in the Gas Furnace section) follows an original method I developed for clocking a pilot. Even manufacturers won't pinpoint how much gas a pilot uses, saying only that it varies considerably. I figured out a way to measure the gas burned by any pilot connected to a meter, and it is covered in Laboratory Worksheet GF-2.

There is some necessary duplication since this laboratory manual may be used in a refrigeration class or in an air conditioning class, but both classes must cover such topics as leak testing, evacuation, and recharging.

I have many people to thank for the successful completion of this laboratory manual. I am grateful to Ed Francis at Pearson/Prentice Hall for giving me the opportunity to share my knowledge and experience with the next generation of HVAC/R technicians. Thanks also to the reviewers of this text for their comments and suggestions: Robert Bates, Delaware Technical and Community College; Arthur Miller, Community College of Allegheny County; and Thomas Niessen, Gateway Technical College. Thanks also to my copyeditor Carol Mohr, who has an eagle eye for spotting errors of all kinds.

I wish a manual like this was available when I started teaching more than 20 years ago. Fortunately for you, now one is. I hope you enjoy your introduction to the career of the HVAC/R technician.

CONTENTS

GAS FURNACE LABORATORIES 141

OIL FURNACE LABORATORIES 187

ELECTRIC FURNACE LABORATORIES 217

REFRIGERATION LABORATORIES 223

GAS HEATING BOILER LABORATORIES   309

AIR CONDITIONING LABORATORIES

BASIC SPRING AC STARTUP

STUDY MATERIAL
Chapter 5, Unit 1

LABORATORY NOTES
One of the most routine calls for the AC service technician is to check out and start up the AC system in the spring. This laboratory worksheet will lead the student through this process.

UNIT DATA

1. Make _____ Model # _____ Serial # _____

2. System type (circle one):   Split system   Package cooling

3. System refrigerant: Type _____ Amount _____

4. Condenser type (circle one):   Air   Water

5. Evaporator coil (circle one): A coil   Slant   H coil   Z coil

6. Below, circle the type of motor for the compressor, condenser, and evaporator. Then write the electrical data from the motor on the voltage (V_{AC}), rated load amps (RLA), and locked rotor amps (LRA) in the spaces provided.

	V_{AC}	RLA	LRA
Compressor (circle one): Permanent split capacitor Capacitor start/capacitor run Three phase	—	—	—
Condenser fan (circle one): Permanent split capacitor Shaded pole	—	—	—
Evaporator fan (circle one): Permanent split capacitor Shaded pole Split phase	—	—	—

3

PRESTART CHECK

Check	Step	Procedure
_____	1.	Manually turn all fan blades. Fans should rotate freely.
_____	2.	Inspect main electrical connections for tightness, signs of overheating, or deterioration of any kind.
_____	3.	Inspect capacitors, circuit board and other wiring terminals for burn marks or deterioration.
_____	4.	Lubricate all motors as required.
_____	5.	High side service valve type (circle one): Service Schrader
_____	6.	Low side service valve type (circle one): Service Schrader
_____	7.	Install gauges and record system idle pressure, that is, the pressure when the system has not been started yet. High side pressure = _____ Low side pressure = _____
_____	8.	Inspect system for any visible signs of leaks, oil, etc.

SYSTEM STARTUP

Check	Step	Procedure
_____	1.	Turn thermostat to a call for cooling.
_____	2.	Install ammeter on C terminal of compressor. On three phase compressors, any lead to compressor will do.
_____	3.	Observe compressor start, draw start amps, amps drop to normal running amps.
_____	4.	Observe/feel normal airflow from condenser.
_____	5.	Observe/feel normal airflow from evaporator.
_____	6.	Allow system 5 min of run time to stabilize pressures. Then record the pressures as follows: High side pressure = _____ Low side pressure = _____
_____	7.	Feel suction line 1 ft from compressor (it should be cool to the touch).
_____	8.	Feel liquid line 1 ft out of condenser (it should be warm to the touch).

SYSTEM MEASUREMENTS

Step	Procedure		
1.	Measure amp draw.		
	Compressor	Voltage = _____	Amperage = _____
	Condenser fan motor	Voltage = _____	Amperage = _____
	Evaporator (indoor) fan motor (IFM)	Voltage = _____	Amperage = _____
2.	Final system operating pressures.	High side pressure = _____	Low side pressure = _____
3.	Describe sight glass condition (TXV systems only). _____		

Step	Procedure

4. Condenser temperature check.

   Temperature of air entering condenser = _____

   Temperature of air leaving condenser = _____

   Temperature rise across condenser = _____

5. Evaporator temperature check.

   Temperature of air entering evaporator = _____

   Temperature of air leaving evaporator = _____

   Temperature difference across evaporator = _____

6. Final system evaluation.

   Comments. _____

   _____

COMPLETE AC STARTUP

STUDY MATERIAL
Chapter 5, Unit 5

LABORATORY NOTES
This laboratory worksheet covers everything the student will need to do for a complete AC startup.

UNIT DATA

1. Unit description _____

2. System type (circle one):   Split system   Package cooling

3. System refrigerant: Type _____   Amount _____

4. Condenser type (circle one):   Air   Water

5. Evaporator coil type (circle one):   A coil   Slant   H coil   Z coil

6. Metering device type (circle one):   TXV   Capillary tube   Orifice

7. Electrical data (ratings):

   Unit   Voltage = _____   Amperage = _____

   Compressor   Voltage = _____   Amperage = _____

   Condenser fan motor   Voltage = _____   Amperage = _____

   Evaporator (indoor) fan motor (IFM)   Voltage = _____   Amperage = _____

PRESTART CHECK

Check	Step	Procedure
_____	1.	Fans rotate freely.
_____	2.	Electrical connections made and checked for tightness.
_____	3.	List all motors requiring field lubrication and the type of lube.
_____	4.	Record belt size by number. Belt size = _____ Belt # = _____

Check	Step	Procedure
_____	5.	Note belt condition and tension. _____
		_____
_____	6.	High side service valve type (circle one): Service Schrader
_____	7.	Low side service valve type (circle one): Service Schrader
_____	8.	All required panels in place or available.
_____	9.	Install gauges. Use red for high side; use blue for low side.
_____	10.	Record system idle pressures. High side pressure = _____ Low side pressure = _____
_____	11.	Note any visible signs of leaks, oil, physical damage, etc. _____
		_____
_____	12.	Capacitor visual inspection. Check capacitor start and capacitor run.

START UP

Check	Step	Procedure
_____	1.	Crankcase heater energized for 8 hr minimum or use the jog-start procedure. For the jog-start procedure, start compressor by depressing contactor and run for 2 sec. Turn off for 1 min. Let oil settle. Repeat three times.
_____	2.	Observe system startup and attain normal system operating pressures.
_____	3.	Measure amperage of motors and determine if actual amperage is within rating.
		Compressor amperage = _____
		Compressor fan motor amperage = _____
		Indoor fan motor amperage = _____
_____	4.	Condenser temperature check.
		Temperature of air entering condenser = _____
		Temperature of air leaving condenser = _____
		Temperature rise across condenser = _____
_____	5.	Evaporator temperature check.
		Temperature of air entering evaporator = _____
		Temperaure of air leaving evaporator = _____
		Temperature difference across evaporator = _____
_____	6.	Install thermometer leads on liquid line 1 ft downstream from condenser and on suction line 1 ft upstream from compressor.

Check	Step	Procedure

_____   7.   Check charge.

     _____   a.   For a TXV system, check either the subcooling or the sight glass. A range of 12–18°F of subcooling is good for TXV.

         TXV: High side pressure = _____

               Low side pressure = _____

               Sight Glass Condition _____

               Liquid line temperature = _____

               Subcooling = _____

               Charge OK? _____

     _____   b.   For a capillary tube/fixed bore system, check the charge with the superheat method. Obtain recommended superheat for today's conditions from superheat calculator or manufacturer charging charts. A value of 20°F +/–5°F superheat is a good average.

         Recommended superheat = _____

         Actual suction line temperature – Coil saturation temperature = Superheat

         _____ – _____ = _____

_____   8.   Use high, low, or OK to evaluate system performance for the following items:

         High side pressure _____   Low side pressure _____   Superheat _____   Subcooling _____

_____   9.   Estimate total system charge (circle one):   Overcharged   Undercharged   OK

_____   10.   Check vibration and noise. Comments: _____

_____   11.   Final system evaluation. Comments: _____

_____   12.   Repairs or changes made or recommended: _____

PREVENTIVE MAINTENANCE

STUDY MATERIAL
Chapter 5, Unit 5

LABORATORY NOTES

Preventive maintenance is an important function. Some customers may think that preventive maintenance is just a way to waste time at their expense. Customer education is therefore very important. If you find an excessively dirty blower, showing this to the customer may make them appreciate the service you are doing more. If you find a partial shorted winding, take the time to explain to the customer how your fixing it will save them money and make their system more efficient. This communication can also develop customer loyalty. Your employer will always appreciate technicians who can communicate well with customers, and doing preventive maintenance is a great opportunity for you to practice this important skill.

UNIT DATA

1. Unit description _____

2. Split system _____   Package cooling only _____

3. System refrigerant: Type _____   Amount _____

4. Condenser type (circle one):   Air   Water

5. Evaporator coil type (circle one):   A coil   Slant   H coil   Z coil

6. Metering device type (circle one):   TXV   Capillary tube   Orifice

7. Electrical data (ratings):

   Unit   Voltage = _____   Amperage = _____

   Compressor   Voltage = _____   Amperage = _____

   Condenser fan motor   Voltage = _____   Amperage = _____

   Evaporator (indoor) fan motor (IFM)   Voltage = _____   Amperage = _____

PRESTART CHECK

Check	Step	Procedure
_____	1.	All fans rotate freely.
_____	2.	All electrical connections checked for tightness.
_____	3.	Lubricate all motors as required.
_____	4.	Record belt size by condition and tension. _____
_____	5.	High side service valve type (circle one): Service Schrader
_____	6.	Low side service valve type (circle one): Service Schrader
_____	7.	All required panels in place or available.
_____	8.	Install gauges. Use red for high side pressure and use blue for low side pressure.
_____	9.	Record system idle pressures. High side pressure = _____ Low side pressure = _____
_____	10.	Note any visible signs of leaks, oil, physical damage, etc. _____
_____	11.	Capacitor visual inspection. Check capacitor start and capacitor run.

MINI STARTUP (5 MINUTE RUN TIME)

Check	Step	Procedure
_____	1.	Crankcase heater energized for 8 hr minimum (or use jog-start procedure).
_____	2.	Observe system start up and attain normal system operating pressures.
_____	3.	Measure amperage of motors and determine if actual amperage is within rating.

Measured amps (circle one):   OK   Not OK

Compressor amps (circle one):   OK   Not OK

Compressor fan motor amps (circle one):   OK   Not OK

Indoor fan motor amps (circle one):   OK   Not OK

| _____ | 4. | Condenser temperature check. |

Temperature of air entering condenser = _____

Temperature of air leaving condenser = _____

Temperature rise across condenser = _____

Check	Step	Procedure
_____	5.	Evaporator temperature check.

Air entering evaporator temperature.

Wet bulb temperature = _____

Dry bulb temperature = _____

Temperature of air leaving evaporator = _____

Temperature difference across evaporator = _____

CLEAN UNIT

Check	Step	Procedure
_____	1.	Turn off, lock out, and tag out at disconnect.
_____	2.	[Optional] Obtain water supply at hose and cleaning agent.
_____	3.	Turn system off at main disconnect.
_____	4.	Remove all required panels.
_____	5.	Remove indoor blower motor assembly.
_____	6.	Wet down condenser, evaporator, blower, and fan blades. Do not spray water directly on any electrical components or motors.
_____	7.	Apply cleaning agent (if used). Some companies clean condensers with plain water.
_____	8.	Using hose or power washer, thoroughly clean all components with enough water to ensure removal of any soap residue left over from the cleaning agent.
_____	9.	Allow minimum of 1/2 hr drain time before reassembly.
_____	10.	Obtain and install new belts and sheaves as required.
_____	11.	Is the gauge manifold still installed? Reinstall if the manifold has been removed.
_____	12.	Reinstall all panels in their appropriate places.
_____	13.	Reinstall thermometer leads on the liquid line 1 ft downstream from the condenser and for the suction line 6 in upstream from the compressor.

FINAL STARTUP, TEMPERATURE, AND PERFORMANCE CHECK

Check	Step	Procedure
_____	1.	Remove lockout tags and lock, turn on main disconnect, and obtain a call for cooling.
_____	2.	Observe system startup and attain normal system operating pressures.

Check	Step	Procedure

_____ 3. Measure amperage of motors and determine if actual amperage is within rating:

        Measured amps (circle one):   OK   Not OK

        Compressor amps (circle one: OK   Not OK

        Compressor fan motor amps (circle one):   OK   Not OK

        Indoor fan motor amps (circle one):   OK   Not OK

_____ 4. Condenser temperature check.

        Air off condenser – Outside air temperature = Temperature rise

        _____ – _____ = _____

_____ 5. Evaporator temperature check.

        Return air dry bulb temperature – Supply air dry bulb temperature

        = Temperature difference across evaporator

        _____ – _____ = _____

        [Optional]

        Return air wet bulb temperature – Supply air wet bulb temperature = Drop in wet bulb temperature

        _____ – _____ = _____

_____ 6. Install thermometer leads on liquid line and suction line.

_____ 7. Measure and record:

High side pressure = _____   Liquid line temperature = _____   Subcooling = _____

_____ 8. Measure and record:

Low side pressure = _____   Suction line temperature = _____   Superheat = _____

_____ 9. Add or remove charge as required to obtain correct operation. Refrigerant added/removed = ___oz

_____ 10. Note any additional required repairs or recommendations. _____

_____

_____ 11. Final system evaluation. _____

_____

REFRIGERANT RECOVERY

STUDY MATERIAL
Chapter 6, Unit 3, Table 6-3-1

LABORATORY NOTES
This lab is intended to help students practice the removal of refrigerant from a system. This is a job that must be done on any and all systems prior to any refrigerant side repairs or dismantling. In the labs that follow recovery will be listed and required on all jobs performing any major refrigerant service. This first time a student is performing a refrigerant recovery it is a job in itself. In future labs and jobs, recovery will be in addition to the bigger job of making the repairs, instead of a job in itself.

UNIT DATA

1. Shop ID #_____

2. Unit description _____

3. System refrigerant: Type _____   Amount _____

4. Factory test pressures: High side pressure = _____   Low side pressure = _____

5. Determine target system pressure from Table 6-3-1 on page 327 of *Refrigeration and Air Conditioning,* 4E.

   Pressure = _____

RECOVER EXISTING REFRIGERANT, BASIC METHOD

Check	Step	Procedure
_____	1.	Install gauges and record system idle pressures.
		High side pressure = _____ Low side pressure = _____
_____	2.	Obtain recovery station, accurate scale, target cylinder partially filled with system refrigerant type and one extra hose.
_____	3.	Weigh the target tank and check refrigerant type. Weight = _____ Type = _____
_____	4.	Read and record tare weight listed on tank. Tare weight = _____

Check	Step	Procedure
_____	5.	Subtract #4 from #3 to calculate the weight of the refrigerant in the tank.
		Weight of refrigerant = _____
_____	6.	Connect gauge manifold center hose to the inlet of the recovery station, that is, the vapor side valve.
_____	7.	Connect extra hose from recovery station to the vapor side of the target tank.
_____	8.	Open low side pressure gauge handle and purge the center hose at the recovery station as required.
_____	9.	Open recovery station outlet valve and purge the hose into the target tank.
_____	10.	Plug in recovery station and turn on pump.
_____	11.	Observe recovery outlet gauge going up.
_____	12.	Open target tank inlet valve.
_____	13.	Observe gauge manifold low side pressure dropping.
_____	14.	Hear and feel refrigerant entering the target tank.
_____	15.	Observe the target tank increase in weight as refrigerant enters the tank.
_____	16.	Do not exceed listed max weight of tank.
_____	17.	Open gauge manifold high side gauge handle to continue recovery from the high side of the system.
_____	18.	Continue recovery process until desired system pressure is reached. Refer to Table 6-3-1 on page 327 of *Refrigeration and Air Conditioning,* 4E.
_____	19.	Close both gauge handles, turn off recovery station, and observe system pressure.
_____	20.	Turn on recovery station to bring the system pressure back down as required by Table 6-3-1.

LEAK TEST AND EVACUATION

STUDY MATERIAL
Chapter 2, Unit 3

LABORATORY NOTES

This laboratory worksheet is intended to allow you to practice on the three basic leak testing and evacuation procedures commonly used in the field: the timed or one hour method, the triple evacuation or triple vac method, and the micron evacuation. Since it is not the intention to recharge the system, your instructor may set up an unrepaired leak for you to find. On such a system it would be impossible to pass a micron evacuation or any vacuum pressure drop test.

When evacuating any system using Schrader valves, the cores are sometimes removed. The purpose is to get a better quality evacuation faster. Some service technicians or instructors may not want to bother with this practice. If the cores are removed, they must be put back in. Some people feel that the time and effort required to put the cores back in and the danger of getting contamination back into the system in the process make removing the cores an ineffective practice.

I have noted the time at which the cores should be removed and put back in but this is optional. Consult your instructor before removing Schrader cores, and put them back in if you take them out.

UNIT DATA

1. Shop ID #_____

2. Unit description _____

3 System application and refrigerant type _____

4. Factory test pressures: High side pressure = _____   Low side pressure = _____

RECOVER EXISTING REFRIGERANT

Refer to Laboratory Worksheet AC-4 for the procedure.

Part 1: LEAK TEST PROCEDURE

Check	Step	Procedure
_____	1.	Install gauges and record system idle pressures.
		High side pressure = _____ Low side pressure = _____
_____	2.	Inspect system for signs of oil, soap, nonoriginal soldering, metal deterioration, or any signs of leak problems.
_____	3.	Obtain halide, electronic, and soap bubbles testers.
_____	4.	Verify correct operation of both halide and electronic testers with a reference leak.
_____	5.	Use existing charge (if any) for first stage of leak testing. If there is no existing charge, then proceed to second stage.
_____	6.	Recover existing charge as required. Check with your instructor.
_____	7.	Pressurize system to a minimum of 30 lb of used R-22 vapor.
_____	8.	Perform a quick leak test with halide and electronic testers.
_____	9.	Observe pressure maintained at 30 lb for 10 min minimum.
_____	10.	Boost the system pressure to 100 lb with nitrogen, helium, or CO_2 for additional leak testing. If necessary, obtain a demonstration on the use of the nitrogen cylinder.
_____	11.	Leak test entire system with both halide and electronic testers.
_____	12.	Use soap bubbles tester to pinpoint any suspected leaks.
_____	13.	Note all leak locations. Consult your instructor before making any repairs. Because good leaks are hard to find, your instructor may want you to demonstrate your leak to other students.
_____	14.	System certified leak free and ready for evacuation.

Part 2: EVACUATION METHOD 1: TIMED METHOD, ONE HOUR

The assumption is made that one hour of evacuation will remove system contamination without actually measuring this removal of contamination. This method is still used my many service mechanics. The time is sometimes varied to fit time available. An evacuation of only 1/2 hr might be good enough if the system was known to be clean. One hour or longer would be even better regardless of system problems. For moisture removal processes, overnight evacuations are common.

Check	Step	Procedure
_____	1.	Bleed leak testing mixture to 0 PSIG.
_____	2.	[Optional] Remove Schrader cores (as required), or move service valve stems to the intermediate position.
_____	3.	Connect vacuum pump to center hose of manifold.
_____	4.	Close both gauge handles.
_____	5.	Turn on vacuum pump.

Check	Step	Procedure
_____	6.	Open both gauge handles slowly. This will prevent excess vapor flow through vacuum pump from blowing oil out the discharge. This step is more important on larger systems that still have pressure in them.
_____	7.	Open gauge handles completely.
_____	8.	Observe low pressure gauge going down and approaching 30 in Hg vacuum.
_____	9.	Record the time under the vacuum pump at which the system reaches 30 in Hg. Time = _____
_____	10.	This is the beginning of your evacuation time. Call your instructor over to verify.

Part 3: EVACUATION METHOD 2: VACUUM PRESSURE DROP TEST (TIMED)

The vacuum pressure drop test, or timed test, is typically done at the conclusion of a timed evacuation or a triple evacuation and used as an additional leak testing procedure. If the system holds 30 in Hg overnight or over a weekend, then there can be no leaks. This is a valid procedure and done as time permits at the discretion of the service technician and under company policy as time permits.

Check	Step	Procedure
_____	1.	Observe evacuation continuing at 30 in Hg and record time.
_____	2.	Close both gauge handles with vacuum pump still operating.
_____	3.	Close blank off valve on vacuum pump.
_____	4.	Remove evacuation hose from pump and connect to blank spud on gauge manifold.
_____	5.	Observe that the system is still at 30 in Hg and record time.
_____	6.	Put vacuum pump away.
_____	7.	Call your instructor over for inspection.

Part 4: EVACUATION METHOD 3: TRIPLE EVACUATION

The triple evacuation procedure means to perform three consecutive evacuations spaced by two dilutions of a dry gas. Nitrogen is preferred but helium and CO_2 can also be used. The clean dry gas will act as a carrier, mixing with system contamination (air and water) and carrying it out during the next evacuation. It is a time consuming procedure but effective in obtaining a clean dry system.

Check	Step	Procedure
_____	1.	Bleed system to 0 PSIG before connecting vacuum pump.
_____	2.	[Optional] Remove Schrader cores (as required), or move service valve stems to the intermediate position.
_____	3.	Begin evacuation and observe pressure dropping observe evacuation reach 30 in Hg.

Check	Step	Procedure
_____	4.	At the conclusion of the first evacuation follow the following steps to obtain a dilution pressure.

    a. Close both gauge handles at the manifold.

    b. Turn off vacuum pump.

    c. Disconnect charging hose from pump.

    d. Install charging hose on the dry gas cylinder pressure regulator output.

    e. Open cylinder and increase pressure output to 50 lb.

    f. Purge air from the center charging hose at the manifold.

    g. Open both gauge handles slowly and obtain the desired dilution pressure

Check	Step	Procedure
_____	5.	Record times for evacuation and dilution time and pressure. Check off each item when you come to it. *Note that the evacuation and dilution pressure can be varied at the discretion of your instructor.*

	Evacuation time	Dilution pressure	Dilution time
1st	___ 30 min ___	___ 10 lb ___	___ 10 min ___
2nd	___ 30 min ___	___ 10 lb ___	___ 10 min ___
3rd	___ 30 min ___		

Part 5: EVACUATION METHOD 4: MICRON EVACUATION

The evacuation pump must be equipped with a blank off valve at the pump. The purpose of the valve is to isolate the pump from the system and yet leave the micron gauge exposed to the system. If the system maintains under 500 microns for a 10 min vacuum pressure drop test, the system is considered clean, dry, and leak free. If a higher pressure of 800 to 1200 microns is maintained, moisture is present in the oil. A leak will cause the micron gauge to rise steadily. A large leak will cause a rapid rise while a small leak will cause a slow rise. Free water in the system will cause a rise to about 20,000 microns.

Check	Step	Procedure
_____	1.	Check vacuum pump.

    _____ a. Obtain good quality vacuum pump and micron pressure gauge.

    _____ b. Change oil in vacuum pump as required.

    _____ c. Connect vacuum pump to micron gauge only.

    _____ d. Check operation of pump and gauge by pulling a vacuum on the sensor.

    _____ e. If vacuum reaches 200 microns or lower it is good.

| _____ | 2. | Micron evacuation. |

    _____ a. Bleed leak test mixture from system to 0 PSIG.

    _____ b. [Optional] Remove Schrader cores (as required), or move service valve stems to the intermediate position.

    _____ c. Install micron pressure gauge in vertical position, preferably at vacuum pump and with a separate valve.

Check	Step	Procedure

_____ d. Begin evacuation and observe pressure dropping.

_____ e. Allow 15 min of evacuation time before exposing vacuum gauge sensor. This will keep oil from depositing on micron pressure sensor.

_____ f. Open valve to micron pressure gauge sensor. Observe pressure (in microns) dropping.

_____ g. Record lowest pressure reached.   Pressure = _____ microns

_____ h. Close both gauge handles to determine quality of evacuation microns pump is now pulling.   Pressure = _____ microns

_____ i. Open gauge handles and continue evacuation of system until system vacuum bottoms out.   Pressure = _____ microns

_____ j. When system reaches approximately 300 microns you are ready to perform a vacuum pressure drop test.

Part 6: EVACUATION METHOD 5: VACUUM PRESSURE DROP TEST

Check	Step	Procedure

_____ 1. Record pressure (in microns) while vacuum is in progress.   Pressure = _____ microns

_____ 2. Close blank off valve leaving system exposed to micron pressure gauge.

_____ 3. Turn off vacuum pump.

_____ 4. Observe micron pressure gauge seek the true level of pressure in the system.

_____ 5. Record microns after 10 min.   Pressure = _____ microns

_____ 6. Use the following criteria to determine system condition.

a. A clean dry system will hold 500 microns or less.

b. A system with some moisture or contamination mixed with the system oil will level out at somewhere between 1000 and 1500 microns.

c. A system with a leak will have first the micron level rise continuously and then the low pressure gauge manifold reading will begin to go up.

d. A system with free water (water not mixed with oil) will climb off the scale on the micron pressure gauge while the compound gauge stays at 30 in Hg.

_____ 7. System passes pressure drop test if it does not climb above 500 microns within 10 min.

_____ 8. Systems not passing can be cleaned up by:

a. Repeat leak testing.

b. Replace filter drier.

c. Change compressor oil.

d. Combine the triple vac with the micron method.

e. Change the oil in the vacuum pump.

f. Retesting the vacuum pump, gauge manifold, and hoses to verify that your equipment will hold 500 microns for the 10 min vacuum pressure drop test.

g. Replace the hoses or the manifold as required.

_____ 9. System certified to have maintained under 500 microns for 10 min.

LEAK TEST, EVACUATE, AND RECHARGE

STUDY MATERIAL
Chapter 2, Unit 3; Chapter 9

LABORATORY NOTES
This laboratory worksheet will follow the 1 hr evacuation and charge by weight method. Residential capillary tube and metering orifice systems are critically charged systems. This means that they take an exact amount of refrigerant. Some typical capillary tube and metering orifice systems include window AC units, split systems, and smaller package equipment. The exact amount of this charge listed in ounces or in pounds and ounces on the nameplate. If the system has been altered this listed charge will be wrong and the system will have to be charged by the superheat method. When the exact weight is listed and correct the charge should still be checked by the superheat method after sufficient run time for stabilization.

UNIT DATA

1. Shop ID # _____

2. Unit description (circle one):   Split system   Mini-split   Package

3. Condenser type (circle one):   Water cooled   Air cooled

4. Evaporator type (circle one):   H coil   A coil   Slant coil

5. System refrigerant: Type _____   Amount _____

LEAK TEST

Check	Step	Procedure
_____	1.	Pull unit plug. System must be off during the leak testing process.
_____	2.	Locate access valves on high and low side of system.
_____	3.	Obtain a gauge manifold with quick seal ends or manual seal ends on all three hoses.
_____	4.	Install gauges and record system idle pressures.
		High side pressure = _____ Low side pressure = _____
_____	5.	Recover existing refrigerant.

Check	Step	Procedure
_____	6.	Install leak testing mixture of 25 PSIG used R-22 vapor as the trace gas and boost to 100 PSIG with nitrogen.
_____	7.	Leak test with halide, soap bubbles, and electronic testers.
_____	8.	Record location of any leaks located.

EVACUATION (ONE HOUR METHOD)

Check	Step	Procedure
_____	1.	Bleed system pressure to 0 PSIG through the center hose. Catch any oil with a rag.
_____	2.	[Optional] Remove Schrader valve cores manually if desired.
_____	3.	Obtain vacuum pump and extension cord.
_____	4.	Check oil level in vacuum pump.
_____	5.	Connect center manifold hose to vacuum pump.
_____	6.	Turn on vacuum pump.
_____	7.	Open both gauge handles on manifold.
_____	8.	Observe compound gauge drop to 30 in Hg.
_____	9.	Allow 1 hr of evacuation time at 30 in Hg to ensure a quality evacuation.
_____	10.	Put away all leak testing equipment and get out charging equipment and supplies during evacuation.
_____	11.	Record total evacuation time.
_____	12.	Close gauge handles on manifold.
_____	13.	Turn vacuum pump off and put away.
_____	14.	Connect center hose to manifold.
_____	15.	Perform vacuum pressure drop test by maintaining 30 in Hg for 48 hr (as desired or time permits).

RECHARGE SYSTEM

Check	Step	Procedure
_____	1.	Observe system holding 30 in Hg vacuum.
_____	2.	Check nameplate data for refrigerant type and amount. _____ oz of R-_____
_____	3.	Obtain an accurate scale (electronic recommended).
_____	4.	Obtain a refrigerant cylinder with sufficient refrigerant to fill system.
_____	5.	Weigh and record the total weight of the cylinder and refrigerant. Weight = _____
_____	6.	Connect center hose to refrigerant cylinder. Use liquid port when charging with any 400 series refrigerant or recovered refrigerant.
_____	7.	Make sure gauge manifold valves are closed.
_____	8.	Open cylinder valve to put pressure on center charging hose.

Check	Step	Procedure
_____	9.	[Optional] If the Schrader valve cores have been removed, this is the time to replace them. To do this, fill the system to 2 PSIG and manually install cores. Do not overtighten.
_____	10.	Open high (red) pressure gauge handle putting refrigerant into high side of system only.
_____	11.	Observe pressure rise on low pressure gauge. You now know that you have normal flow of refrigerant from high to low side of system.
_____	12.	Add refrigerant until correct weight is installed. Refrigerant can be put into low side also once normal flow has been verified.
_____	13.	Close both high and low pressure gauge handles and tank valve.
_____	14.	Turn system on.
_____	15.	Observe high side pressure go up and low side pressure go down.
_____	16.	Drain center hose into low side by opening low side pressure gauge handle. Observe the low side gauge go up and then down.
_____	17.	Close the low side gauge handle.
_____	18.	Allow sufficient run time for system stabilization.

WARM WEATHER SIMULATION TEST

Typically in an HVAC lab and frequently in the real world the startup and system test does not take place on a design day. To observe actual typical operating conditions we must simulate them. We can partially block off the condenser airflow on the inlet side until the high side pressure is 260 lb R-22 or 120°F condensing temperature. The indoor ambient temperature will still be a little low, causing a lower than normal low side pressure, but it is closer to a warm summer day than average indoor shop conditions.

FINAL TEMPERATURE PRESSURE CHECK

Check	Step	Procedure
_____	1.	With condenser partially blocked off, measure and record high side pressure and low side pressure.
_____	2.	Measure and record suction line temperature and liquid line temperature.
_____	3.	Calculate system operating superheat by the following formula:

Suction line temperature – Low side pressure coil temperature = Superheat

_____ – _____ = _____

| _____ | 4. | Calculate system operating subcooling by the following formula: |

High side coil temperature – Liquid line temperature = Subcooling

_____ – _____ = _____

| _____ | 5. | Demonstrate system operation for your instructor. |

LEAK TEST, EVACUATE, AND RECHARGE

STUDY MATERIAL
Chapter 9

LABORATORY NOTES
Charge by Superheat (capillary tube only) Triple evacuation

This lab will use the charge by superheat method for capillary tube systems and the triple evacuation (triple vac) procedure of three consecutive evacuations spaced by two dilutions of an inert gas to obtain a clean dry system. The times can be varied to fit the time available. This standard method was considered obsolete by some people in the industry with the invention of the modern two stage rotary deep vacuum pump many years ago. But the triple vac is still a reliable method of obtaining a clean dry system and is recommended by some manufacturers.

The system is charged by weight using an electronic scale to weigh refrigerant. Check charge by measuring operating pressures, superheat, and subcooling. Record the total weight of the final charge.

UNIT DATA

1. Shop ID #_____

2. Unit description _____

3. System application and refrigerant type: _____

4. Factory test pressures: High side pressure = _____   Low side pressure = _____

LEAK TEST PROCEDURE

Check	Step	Procedure
_____	1.	Install gauges, red on high side and blue on low side and record system idle pressures.
		High side pressure = _____ Low side pressure = _____
		Note that if system idle pressures are saturated for refrigerant type R-22 or 410A, assume no major leaks and proceed to step 4.
_____	2.	Pressurize system to a minimum of 35 PSIG R-22 for the first round of leak testing. Record leak locations if found.

Check	Step	Procedure
_____	3.	Boost the system pressure to 100 lb with the nitrogen, helium, or CO_2 for additional leak testing. If necessary, obtain a demonstration on the use of the nitrogen cylinder.
_____	4.	Use the halide, electronic, and soap bubbles testers to become familiar with each type of device.
_____	5.	Record all leak locations and consult your instructor before making any repairs. Remember the saying that good leaks are hard to find.
_____	6.	System certified leak free and ready for evacuation.

EVACUATION

The triple evacuation procedure means to perform three consecutive evacuations spaced by two dilutions of a dry gas. Nitrogen is preferred but helium and CO_2 can also be used. The clean dry gas will act as a carrier, mixing with system contamination (air and water) and carrying it out during the next evacuation. It is a time consuming procedure but effective in obtaining a clean dry system.

Check	Step	Procedure
_____	1.	Bleed system to 0 PSIG before connecting vacuum pump.
_____	2.	[Optional] Remove the Schrader valve cores (as required) or put the full service valves in intermediate position. Note that cores are removed to get a faster and better quality evacuation. If you take them out they must be put back in later. Some technicians choose to leave them in and evacuate a little longer. A micron pressure gauge can be used to measure the quality of evacuation, but not all technicians have them or choose to use them. Laboratory Worksheets AC-5 and AC-8 give the procedure using a micron pressure gauge.
_____	3.	Install gauges on both sides of system, red to high and blue to low pressure side of system.
_____	4.	Begin evacuation and observe pressure dropping. Observe evacuation reach 30 in Hg.
_____	5.	At the conclusion of the first evacuation, complete the following steps to obtain a dilution pressure.
		_____ a. Close both gauge handles at the manifold.
		_____ b. Turn off vacuum pump, disconnect hose from pump.
		_____ c. Install charging hose on the dry gas cylinder pressure regulator output.
		_____ d. Open cylinder and increase pressure output to 50 lb.
		_____ e. Purge air from the center charging hose at the manifold.
		_____ f. Open both gauge handles slowly and obtain the desired dilution pressure.
_____	6.	Record times for evacuation and dilution time and pressure. Check off each measurement as you make it. Evacuation and dilution pressure can be varied at the discretion of your instructor.

	Evacuation time	Dilution pressure	Dilution time
1st	___30 min___	___10 lb ___	___10 min ___
2nd	___30 min___	___10 lb ___	___10 min ___
3rd	___30 min___		

RECHARGE PROCEDURE

Weigh in the charge and measure the superheat to verify correct operation.

Check	Step	Procedure
_____	1.	Obtain an accurate electronic scale.
_____	2.	Obtain a cylinder with sufficient refrigerant.
_____	3.	Record the weight of the refrigerant in the cylinder.
_____	4.	Add refrigerant as required to obtain a positive pressure.
_____	5.	[Optional] Reinstall Schrader valve cores (if they have been removed) with the system pressures balanced at 2 PSIG.
_____	6.	Open the high side gauge handle and put liquid into the high side of the system until the system high side equals the cylinder pressure.
_____	7.	Close the high pressure gauge handle and wait until the system pressures are equalized.
_____	8.	Turn the system on and observe the high pressure gauge go up and the low pressure gauge go down.
_____	9.	Add additional refrigerant vapor (turn cylinder upright for vapor) into the low side of the operating system until normal operating pressures are observed. Typical in shop values are as follows, high side pressure = 225 PSIG and low side pressure = 60 PSIG. See your instructor for other system pressures.
_____	10.	Install a remote lead thermometer on the suction line 12 in upstream from the compressor.
_____	11.	Install condenser air blocking with metal, paper, etc., as required to raise the high side pressure to simulate summer conditions.
_____	12.	Add refrigerant (vapor into low side) as required to obtain a high side pressure value of 265 PSIG and a low side pressure of 65 PSIG with 20°F of suction line superheat.
_____	13.	Record total weight of refrigerant installed.
_____	14.	Call for instructor inspection at this point.

LEAK TEST, EVACUATE, AND RECHARGE

STUDY MATERIAL
Chapter 2, Unit 3; Chapter 9

LABORATORY NOTES

This laboratory worksheet will cover the process of micron pressure gauge evacuation and charge by sight glass. This version of the leak test, evacuate, and recharge uses the micron gauge to measure the evacuation and the system is to be charged using the sight glass. The system must be clean, dry, and leak free to pass the 500 micron vacuum pressure drop test. It is recommended to install a new filter drier before beginning the evacuation. By measuring the vacuum in microns we can detect leaks and determine the presence of moisture in the system.

The vacuum pump must be equipped with a blank off valve. The purpose of the valve is to isolate the pump from the system and leave the micron gauge exposed to the system. If the system maintains under 500 microns for a 10 min vacuum pressure drop test, the system is considered clean, dry, and leak free. If 800 to 1200 microns is maintained, moisture is present in the oil. A leak will cause the micron gauge to rise steadily, a large leak will cause a fast rise rapidly while a small leak will cause a slow rise. Free water in the system will cause a rise to about 20,000 microns.

UNIT DATA

1. Shop ID # _____

2. Unit description _____

3. System application and refrigerant type: _____

4. Factory test pressures: High side pressure = _____   Low side pressure =_____

LEAK TEST PROCEDURE

Check	Step	Procedure
_____	1.	Install gauges and record system idle pressures.
		High side pressure = _____ Low side pressure = _____
_____	2.	Pressurize system to 25 lb R-22 for first pressure test.
_____	3.	If no leaks can be found at 25 lb, then boost pressure to 100 lb with nitrogen, helium, or CO_2.

Check	Step	Procedure
_____	4.	Obtain demonstration (as required) on use of nitrogen cylinder and pressure regulator. To pressure test at higher pressures, do not exceed factory test pressures.
_____	5.	Use the halide, soap bubbles, and the electronic leak testers sufficient to become familiar with each.
_____	6.	Record all leak locations and demonstrate to your instructor. Do not automatically repair a leak. Your instructor may want to show other students your leak, due to the saying that good leaks are hard to find.
_____	7.	Use nitrogen to pressure test at higher pressures. *Note: Do not exceed factory test pressures.*
_____	8.	Repeat as required.
_____	9.	System certified, all leaks found, and system ready for evacuation.

CHECK VACUUM PUMP

Check	Step	Procedure
_____	1.	Obtain good quality vacuum pump and micron pressure gauge.
_____	2.	Change oil in vacuum pump as required.
_____	3.	Connect vacuum pump to micron pressure gauge only.
_____	4.	Check operation of pump and gauge by pulling a vacuum on sensor.
_____	5.	If vacuum reaches 200 microns or lower it is OK.

EVACUATION

Check	Step	Procedure
_____	1.	Bleed leak test mixture from system to 0 PSIG.
_____	2.	[Optional] Remove Schrader valve cores or put service valves to intermediate position. If you remove the Schrader cores, you must put them back in before the system is charged.
_____	3.	Install gauges on both sides.
_____	4.	Install micron gauge in vertical position preferably at the vacuum pump and with a separate valve.
_____	5.	Begin evacuation and observe pressure dropping.
_____	6.	Allow 15 min of evacuation time before exposing vacuum gauge sensor. This will keep oil from depositing on micron pressure sensor.
_____	7.	Open valve to micron gauge sensor and observe pressure dropping.
_____	8.	Record lowest pressure reached. Pressure = _____ microns
_____	9.	Close both gauge handles to determine quality of evacuation vacuum pump is now pulling.
_____	10.	Open gauge handles and continue evacuation of system until system vacuum bottoms out.
_____	11.	Close blank off valve leaving system exposed to micron pressure gauge. This is the vacuum pressure drop test.

Check	Step	Procedure
_____	12.	Record pressure after 10 min. Pressure = _____ microns
_____	13.	System passes pressure drop test if it does not climb above 500 microns within 10 min.
_____	14.	System certified to have maintained under 500 microns for 10 min. A value of 1200 microns or less indicates moisture in the oil. Continuous climbing pressure indicates a leak or free water.

RECHARGE PROCEDURE

Use sight glass method for TXV systems only. Use charge by superheat or weight for capillary tube systems.

Check	Step	Procedure
_____	1.	Close both gauge handles.
_____	2.	Remove the sensor, gauge, and vacuum pump, in that order.
_____	3.	[Optional] If Schrader cores have been removed, fill system to 2 PSIG using system refrigerant.
_____	4.	[Optional] Reinstall the cores of a Schrader valve system at this time.
_____	5.	Fill the system with cylinder vapor pressure.
_____	6.	Close the high side gauge handle of the manifold.
_____	7.	Start the system up, observe the high side pressure going up and the low side pressure going down.
_____	8.	Add refrigerant into the low side during operation until the sight glass begins to show some liquid.
_____	9.	Close the low side gauge handle, stopping the flow of refrigerant into the system, and observe system operation.
_____	10.	Add refrigerant occasionally to clear the sight glass of bubbles.
_____	11.	In the absence of a sight glass, 12–15°F of subcooling can be considered a full charge.

FINAL SYSTEM TEMPERATURE AND PRESSURE CHECK

Check	Step	Procedure
_____	1.	Record the following values:

        High side pressure = _____

        Low side pressure = _____

        Ambient temperature = _____

        Suction line temperature = _____

        Liquid line temperature = _____

        System subcooling = _____

CHECK CHARGE BY MEASURING SUPERHEAT

STUDY MATERIAL
Chapter 9, Unit 2

LABORATORY NOTES

This laboratory worksheet uses the R-22 refrigerant in a capillary tube system. There are many AC systems in use that have very minor leak problems that may or may not be located or repairable. Before installing refrigerant in any system we must assume there is a leak and attempt to find the leak or leaks and repair them. If we find a leak that can be repaired by tightening a fitting or installing a new capillary tube, a complete recovery is not necessary. This job is to do a mini-startup, leak test, repair leaks as required, and add refrigerant to a full charge by measuring superheat.

UNIT DATA

1. Shop ID # _____

2. Unit make _____   System capacity _____

3. Refrigerant type: <u>R-22</u>

4. Metering device: <u>Capillary tube</u>

PRESTART CHECKS

Check	Step	Procedure
_____	1.	Inspect the system to be sure it is a capillary or fixed orifice system.
_____	2.	All fans turn freely.
_____	3.	System installation complete and ready to run.
_____	4.	Thermostat installed and operable.
_____	5.	Install gauges and record system idle pressures.

High side pressure = _____   Low side pressure = _____

DETERMINE NORMAL OPERATING PRESSURES

Check	Step	Procedure
_____	1.	Use manufacturer charging charts to determine the expected normal operating pressures.

High side pressure = _____   Low side pressure = _____

CHECK SYSTEM OPERATION

Check	Step	Procedure
_____	1.	If system is out of refrigerant, proceed to leak test, evacuate, and recharge.
_____	2.	If system is low on refrigerant but not out, system may be leak tested and recharged without recovery.
_____	3.	Start up system by turning thermostat to a setting lower than the room temperature, and set the heat-off-cool switch to cool.
_____	4.	Observe/verify normal airflow across condenser and evaporator.
_____	5.	Observe that the compressor is on by noting that the high side pressure is climbing and the low side pressure is dropping.
_____	6.	Install suction line thermostat 1 ft upstream from compressor.
_____	7.	Install liquid line thermostat at condenser outlet.
_____	8.	Record system operating pressures after 5 min of system operation.

High side pressure = _____   Low side pressure = _____

| _____ | 9. | Measure suction superheat by the following formula: |

Suction line temperature – Low side saturation temperature = Superheat

_____ – _____ = _____

| _____ | 10. | Measure liquid subcooling by the following formula: |

High side saturation temperature – Liquid line temperature = Subcooling

_____ – _____ = _____

| _____ | 11. | Verify system low on refrigerant by the following criteria. Low side pressure low, high side pressure low, superheat high, and subcooling low. |
| _____ | 12. | Turn system off and observe pressures equalize. |

LEAK TEST

Check	Step	Procedure
_____	1.	Perform leak test with system idle and fans off.
_____	2.	Using halide, soap bubbles, and electronic leak testers, test all system fittings for leaks.
_____	3.	Record location of any leaks found and repaired. _____
		_____
_____	4.	Leaks in mechanical joints that can be repaired by tightening or installing new caps, will not require a system recovery or evacuation.
_____	5.	System certified as ready for partial recharge with all repairable leaks found and repaired.

RECHARGE

Check	Step	Procedure
_____	1.	Using manufacturer charts determine correct operating pressures and superheat for your ambient temperature condition.
		High side pressure = _____ Low side pressure = _____ Superheat = _____
_____	2.	Get help from your instructor if recommended pressures and superheat are not available.
_____	3.	Obtain normal system operation.
_____	4.	Run system 5 min to stabilize operation.
_____	5.	Set refrigerant on accurate scale.
_____	6.	Record original weight of refrigerant cylinder. ___lb ___oz
_____	7.	Add refrigerant a few ounces at a time, vapor into the low side during compressor operation.
_____	8.	Monitor high side pressure, low side pressure, and superheat.
_____	9.	Do not exceed expected high side pressure, low side pressure, or 12°F of subcooling.
_____	10.	Charge until expected superheat and pressures are reached. You will probably not get them exactly. Call your instructor when you get close.

CHECK CHARGE BY MEASURING SUBCOOLING

STUDY MATERIAL
Chapter 9, Unit 2

LABORATORY NOTES

This laboratory worksheet covers an R-22 system using a TXV. There are many AC systems in use that have very minor leak problems that may or may not be located or repairable. Before installing refrigerant in any system we must assume there is a leak and attempt to find the leak or leaks and repair them. If we find a leak that can be repaired by tightening a fitting or installing a new capillary tube, a complete recovery is not necessary. This job is to do a mini-startup, leak test, leak repair, and add refrigerant to a full charge by measuring subcooling. *Note: Measuring super-heat on a TXV system is a test of the TXV and not to be used as the final criterion of a correct charge as in a fixed orifice system.*

UNIT DATA

1. Shop ID # _____

2. Unit make _____

3. Refrigerant type _____   System capacity _____

4. Metering device: <u>TXV</u>

PRESTART CHECKS

Check	Step	Procedure
_____	1.	Inspect the system to be sure it is a TXV system.
_____	2.	All fans turn freely.
_____	3.	System installation complete and ready to run.
_____	4.	Thermostat installed and operable.
_____	5.	Install gauges and record system idle pressures.

High side pressure = _____   Low side pressure = _____

DETERMINE NORMAL OPERATING PRESSURES

Check	Step	Procedure
_____	1.	Use manufacturer charging charts to determine the expected normal operating pressures. High side pressure = _____ Low side pressure = _____

CHECK SYSTEM OPERATION

Check	Step	Procedure
_____	1.	If system is out of refrigerant proceed to leak test, evacuate, and recharge.
_____	2.	If system is low on refrigerant but not out, system may be leak tested and recharged without recovery.
_____	3.	Start up system by turning thermostat to a setting lower than the room temperature, and set the heat-off-cool switch to cool.
_____	4.	Observe/verify normal airflow across condenser and evaporator.
_____	5.	Observe that the compressor is on by noting the high side pressure is climbing and the low side pressure is dropping.
_____	6.	Install suction line thermostat 1 ft upstream from compressor.
_____	7.	Install liquid line thermostat at condenser outlet.
_____	8.	Record system operating pressures after 5 min of system operation. High side pressure = _____ Low side pressure = _____
_____	9.	Measure suction superheat by the following formula: Suction line temperature – Low side pressure saturation temperature = Superheat _____ – _____ = _____
_____	10.	Measure liquid subcooling by the following formula: High side pressure saturation temp – Liquid line temperature = Subcooling _____ – _____ = _____
_____	11.	Verify system low on refrigerant by the following conditions being present. Low side pressure low, high side pressure low, superheat high, and subcooling low.
_____	12.	Turn system off, plug center hose of gauge manifold, and open both gauge handles to equalize system pressures.

LEAK TEST

Check	Step	Procedure
_____	1.	Perform leak test with system idle and fans off.
_____	2.	Using halide, soap bubbles, and electronic leak testers, test all system fittings for leaks.
_____	3.	Record and report the location of any leaks found.
_____	4.	Leaks in mechanical joints that can be repaired by tightening or installing new caps will not require a system recovery or evacuation.
_____	5.	System certified as ready for partial recharge with all repairable leaks found and repaired.

RECHARGE

Check	Step	Procedure
_____	1.	Using manufacturer charts, if available, determine correct operating pressures for your ambient condition. High side pressure = _____ Low side pressure = _____
_____	2.	Get help from your instructor if recommended pressures or charging charts are not available.
_____	3.	Obtain normal system operation.
_____	4.	Run system 5 min to stabilize operation.
_____	5.	Set refrigerant on accurate scale.
_____	6.	Record original weight of refrigerant cylinder. _____lb _____oz
_____	7.	Add refrigerant a few ounces at a time, vapor into the low side during compressor operation.
_____	8.	Monitor high side pressure, low side pressure, and subcooling.
_____	9.	Do not exceed expected high side pressure, low side pressure, or 12°F of subcooling.
_____	10.	Charge till expected subcooling and pressures are reached. You will probably not get them exactly. Call your instructor when you get close.

INSTALL A CENTRAL SPLIT SYSTEM

STUDY MATERIAL
Chapter 9, Unit 1

LABORATORY NOTES

Any central forced air heating can be equipped with an add on cooling system. Some central systems perform much better than others, the differences being in the volume delivered the location of the register outlets, and the size and type of outlets used. The best registers for cooling systems are high sidewall, ceiling, and baseboard, in that order. Most heat only systems move an inadequate air volume for cooling systems. The job of installing an add-on central system includes setting the refrigeration side components, completing the refrigeration piping, performing a leak test, evacuation, and recharge procedure, improving the airflow to 400 CFM per ton of cooling, and to wire the cooling components for main power and control circuit.

LOCATE THE REFRIGERATION COMPONENTS

Check	Step	Procedure
_____	1.	A matched cased coil is really an extension of the furnace. When using a matched cased coil, set it directly on the furnace outlet and proceed to step 3.
_____	2.	Set the evaporator coil within the plenum level and firm on L brackets inserted into existing S cleats. An opening must be cut into the existing plenum to install a coil support ledge inside an existing plenum. Be sure to leave coil access for cleaning when covering this opening.
_____	3.	Set condensing unit outside in an accessible location with plenty of airflow, away from the building, on a separate support pad. Follow manufacturer's recommendation for the airflow of the condenser fan.

RUN LINES

Check	Step	Procedure
_____	1.	Look over building and drill holes for line penetration through building walls, etc.
_____	2.	Install tubing hangers in appropriate places.
_____	3.	Run line set and support on hangers.
_____	4.	Connect ends with provided threaded connections or make soldered connections.

LEAK TEST, EVACUATE, AND CHARGE

Check	Step	Procedure
_____	1.	A system using precharged lines will be completed at this point if the charge is not lost.
_____	2.	Even a precharged system should be leak tested at the threaded connections.
_____	3.	A normal leak test, evacuate, and recharge should be followed at this time.

INSTALLATION COMPLETION

Check	Step	Procedure
_____	1.	Install a condensate drain. Use a condensate pump if an adequate drain is not available.
_____	2.	Enclose and seal the plenum chamber for air leaks.
_____	3.	Insulate and tape any gaps in the suction line.
_____	4.	Seal any gaps in the building penetrations.
_____	5.	Check for correct voltage at the main power supply of the condensing unit.
_____	6.	Install and wire the cooling control circuit.
_____	7.	Install larger fan motor and sheave or wire for correct cooling speed of fan motor operation as required.
_____	8.	System certified ready for charging and startup.
_____	9.	Call your instructor for an inspection before running the system.

SYSTEM STARTUP

This procedure is a short form of the normal AC startup procedure job sheet.

Check	Step	Procedure
_____	1.	Install gauges and note system idle pressures.
_____	2.	Turn on main power supply to condensing unit and furnace.
_____	3.	Turn thermostat to a call for cooling position.
_____	4.	Observe the following:
		_____ a. Condensing unit on
		_____ b. Furnace fan on
		_____ c. High side pressure going up
		_____ d. Low side pressure going down
		_____ e. Cool suction line
		_____ f. Warm liquid line

SYSTEM FINAL OPERATING MEASUREMENTS

Check	Step	Procedure

_____   1.   Calculate temperature difference.

Condenser air outlet temperature – Air outlet temperature = Temperature difference

_____ – _____ = _____

_____   2.   Calculate temperature difference.

Evaporator air inlet temperature – Air outlet temperature = Temperature difference

_____ – _____ = _____

_____   3.   System operating pressures.   High side pressure = _____   Low side pressure = _____

_____   4.   Calculate superheat.

Suction line temperature – Evaporator saturation = Suction line superheat

_____ – _____ = _____

_____   5.   Calculate subcooling.

Condenser saturation – Liquid line temperature = Liquid line subcooling

_____ – _____ = _____

REPLACE A WELDED HERMETIC COMPRESSOR

STUDY MATERIAL
Chapters 5 & 9

LABORATORY NOTES

A hermetic compressor replacement is a very common job in the air conditioning and refrigeration service industry. Unlike the semihermetic replacement, a torch is needed due to the brazed connections. Because of the welded hermetic design, the exact cause of failure cannot always be pinpointed. A good rule is to perform a complete check of the system during the startup to prevent repeat failures. Field replacement of a hermetic compressor should include a new filter drier in the liquid line. Any evidence of acid burnout should add a suction filter to the system. Standard refrigeration or air conditioning startup procedures should be followed to determine correct operation after the compressor replacement. It is not recommended to reuse refrigerant in a hermetic compressor replacement.

REPLACE COMPRESSOR

Check	Step	Procedure
_____	1.	Obtain exact replacement, original equipment manufacturer recommendation, or equivalent replacement compressor using physical fit, voltage, capacity, refrigerant type and temperature range as criteria.
_____	2.	Install gauges on system in the normal manner.
_____	3.	Record system idle pressures.
_____	4.	Attempt compressor operation to verify the failure and reason for replacement.
_____	5.	Demonstrate failure mode for customer as required.
_____	6.	Turn compressor off and lockout main disconnect.
_____	7.	Recover refrigerant to 10 in Hg vacuum and vent system. [Optional] Vent system by removing Schrader valve cores.
_____	8.	Desolder suction and discharge lines, or twist off rota-lock fitting if used.
_____	9.	Remove compressor mounting bolts.
_____	10.	Draw a wiring diagram and identify all wires within the compressor terminal box.
_____	11.	Remove all wires entering terminal box.
_____	12.	Lift out compressor from condensing unit.

Check	Step	Procedure
_____	13.	Lift and install replacement compressor.
_____	14.	Start all compressor bolts.
_____	15.	Using appropriate solder type (consult your instructor) solder suction and discharge lines into the compressor.
_____	16.	Install replacement filter.
_____	17.	Pressurize compressor and system on both sides.
_____	18.	Leak test compressor and filter connections and entire system as required.
_____	19.	Evacuate system using recommended evacuation procedure.
_____	20.	Complete wiring connections, replace any system start components, and tighten compressor bolts during evacuation.
_____	21.	Close gauge valves at manifold.
_____	22.	Move service valve stems to cracked off backseat.
_____	23.	Close the valves and perform a vacuum pressure drop test appropriate for evacuation procedure, as time permits.
_____	24.	Install holding charge. [Optional] Replace Schrader valve cores, if cores were removed.
_____	25.	Charge by recommended charging procedure, by weight, sight glass method, or superheat.
_____	26.	Start up system and obtain normal operation.
_____	27.	Record time for complete compressor change and obtain signature.

AIR SOURCE HEAT PUMP STARTUP

STUDY MATERIAL
Chapter 13

LABORATORY NOTES
This laboratory worksheet covers the startup procedure for the air source heat pump.

UNIT DATA

1. Manufacturer _____   Model # _____   Serial # _____

2. System type (circle one):   Split   Package   Air to air

3. System refrigerant type (circle one):   R-22   R-410A

4. Indoor coil shape (circle one):   A coil   Slant   H coil   Z coil

5. Below, circle the type of motor for the compressor, condenser, and evaporator. Then write the electrical data from the motor on the voltage (V_{AC}), rated load amps (RLA), and locked rotor amps (LRA) in the spaces provided.

	V_{AC}	RLA	LRA

Compressor (circle one): Permanent split capacitor   Capacitor start/capacitor run   Three phase

                                                                             —   —   —

Condenser fan (circle one): Permanent split capacitor   Shaded pole   —   —   —

Evaporator fan (circle one): Permanent split capacitor   Shaded pole   Split phase ___   —   —

PRESTART CHECK

Check	Step	Procedure
_____	1.	Manually turn all fan blades. Fans should rotate freely.
_____	2.	Inspect main electrical connections for tightness, signs of overheating, or deterioration of any kind.
_____	3.	Inspect capacitors, circuit board, and other wiring terminals for burn marks other deterioration.
_____	4.	Lubricate all motors as required.

Check	Step	Procedure
_____	5.	High side service valve type (circle one): Service Schrader
_____	6.	Low side service valve type (circle one): Service Schrader
_____	7.	Install gauges and record system idle pressure, that is, the pressure when the system has not been started yet. High side pressure = _____ Low side pressure = _____
_____	8.	Inspect system for any visible signs of leaks, oil, etc.

COOLING SYSTEM STARTUP

Check	Step	Procedure
_____	1.	Turn thermostat to a call for cooling.
_____	2.	Feel warm air coming out of outdoor coil.
_____	3.	Feel cool air coming out of indoor coil.
_____	4.	Allow system 5 min of run time to stabilize pressures and record. High side pressure = _____ Low side pressure = _____
_____	5.	Measure amp draw. Compressor = _____ Outdoor fan motor = _____ Indoor fan motor = _____
_____	6.	Outdoor coil temperature check. Air leaving condenser – Air entering = Temperature difference _____ – _____ = _____
_____	7.	Indoor coil temperature check. Air entering condenser – Air leaving = Temperature difference _____ – _____ =_____
_____	8.	Measure system superheat. Coil temperature – Suction line temperature = Superheat _____ – _____ =_____
_____	9.	Measure system subcooling. Condensing temperature – Liquid line temperature = Subcooling _____ – _____ =_____

HEAT MODE OPERATION

Check	Step	Procedure
_____	1.	Turn thermostat to a call for heat.
_____	2.	Observe/hear reversing valve change position.
_____	3.	Record new operating pressures. High side pressure = _____ Low side pressure = _____
_____	4.	Feel cool air coming out of outdoor coil.
_____	5.	Feel warm air coming out of indoor coil.
_____	6.	Outdoor coil temperature check.

Air entering – Air leaving = Temperature difference

_____ – _____ = _____

| _____ | 7. | Indoor coil temperature check. |

Air leaving – Air entering = Temperature difference

_____ – _____ = _____

| _____ | 8. | Measure system superheat. |

Coil temperature – Suction line temperature = Superheat

_____ – _____ = _____

| _____ | 9. | Measure system subcooling. |

Condensing temperature – Liquid line temperature = Subcooling

_____ – _____ = _____

WATER SOURCE HEAT PUMP STARTUP

STUDY MATERIAL
Chapter 13

LABORATORY NOTES
This laboratory worksheet covers the startup procedure for a water source heat pump.

UNIT DATA

1. Make _____ Model # _____ Serial # _____

2. System type (circle one): Split Package Water source X

3. System refrigerant type (circle one): R-22 R-410A

4. Indoor coil shape (circle one): A coil Slant H coil Z coil

5. Below, circle the type of motor for the compressor, condenser, and evaporator. Then write the electrical data from the motor on the voltage (V_{AC}), rated load amps (RLA), and locked rotor amps (LRA) in the spaces provided.

	V_{AC}	RLA	LRA
Compressor (circle one): Permanent split capacitor Capacitor start/capacitor run Three phase	—	—	—
Condenser fan (circle one): Permanent split capacitor Shaded pole	—	—	—
Evaporator fan (circle one): Permanent split capacitor Shaded pole Split phase	—	—	—

PRESTART CHECK

Check	Step	Procedure
_____	1.	Manually turn all fan blades. Fans should rotate freely.
_____	2.	Inspect main electrical connections for tightness, signs of overheating, or deterioration of any kind.
_____	3.	Inspect capacitors, circuit board, and other wiring terminals for burn marks or deterioration.

Check	Step	Procedure
_____	4.	Lubricate all motors as required.
_____	5.	High side service valve type (circle one): Service Schrader
_____	6.	Low side service valve type (circle one): Service Schrader
_____	7.	Install gauges and record system idle pressure. System has not been started yet. High side pressure = _____ Low side pressure = _____
_____	8.	Inspect system for any visible signs of leaks, oil, etc.

COOLING MODE SYSTEM STARTUP

Check	Step	Procedure
_____	1.	Turn thermostat to a call for cooling.
_____	2.	Allow system 5 min of run time to stabilize pressures and record. High side pressure = _____ Low side pressure = _____
_____	3.	Feel warm water coming out of water coil.
_____	4.	Feel cool air coming out of indoor coil.
_____	5.	Measure amp draw: Compressor = _____ Outside fan motor = _____ Inside fan motor = _____
_____	6.	Water coil temperature check. Water leaving – Water entering = Temperature difference _____ – _____ = _____
_____	7.	Air coil temperature check. Air entering – Air leaving = Temperature difference _____ – _____ = _____
_____	8.	Measure system superheat. Coil temperature – Suction line temperature = Superheat _____ – _____ = _____
_____	9.	Measure system subcooling. Condensing temperature – Liquid line temperature = Subcooling _____ – _____ = _____

HEAT MODE OPERATION

Check	Step	Procedure
_____	1.	Turn thermostat to a call for heat.
_____	2.	Observe/hear reversing valve change position.
_____	3.	Record new operating pressures. High side pressure = _____ Low side pressure = _____
_____	4.	Feel cool water coming out of water coil.
_____	5.	Feel warm air coming out of indoor coil.
_____	6.	Water coil temperature check.

Water entering – Water leaving = Temperature difference

_____ – _____ = _____

| _____ | 7. | Air coil temperature check. |

Air leaving – Air entering = Temperature difference

_____ – _____ = _____

| _____ | 8. | Measure system superheat. |

Coil temperature – Suction line temperature = Superheat

_____ – _____ = _____

| _____ | 9. | Measure system subcooling. |

Condensing temperature – Liquid line temperature = Subcooling

_____ – _____ = _____

INSTALL A PACKAGE SYSTEM

STUDY MATERIAL
Chapter 17

LABORATORY NOTES

Packaged HVAC equipment systems are generally installed on the roof for bottom air discharge or on a ledge for side discharge. Some units can be switched from bottom discharge to side discharge by moving a panel from the side to the bottom. In a typical school shop setting this is not possible. In our shop, we set bottom discharge units on a platform high enough to get airflow through the unit. Side discharge units will have normal airflow while sitting directly on the floor. A base of some type of vibration absorbing material is a good idea. Most units will require some leveling. Only those units with an electric supply can be put on wheels bolted directly to the unit base or on rolling carts. Units with gas lines require a fixed position. Water drain for condensate can be to a floor drain, a temporary hose connection, or to a movable bucket or pan. Electrical connections to fixed units are typically hardwired from a conduit but in HVAC labs frequently flexible cords and plugs are used.

UNIT DATA

1. Unit name _____   Model # _____   Serial # _____

2. Air discharge (circle one):   Side   Bottom

3. Cooling (circle one):   DX   Chilled water

4. Heating (circle one):   Gas   Electric   Hot water

5. Cooling capacity = _____   Heating capacity = _____

6. Cooling airflow required = _____   Duct size at .01 static = _____

SYSTEM EVALUATION

Place unit in position and briefly describe each of the following.

1. Required airflow in CFM and direction is obtained. _____

   _____

2. Location of and method of connection to a condensate drain. _____

_____

3. Natural gas connection or other fuel line connection. _____

_____

4. List type of electrical connection to be utilized. _____

_____

5. List voltage, phase, amperage, wire size, and number of wires required.

Voltage = _____

Amperage = _____

Wire size = _____

Number of wires = _____

TIME AND MATERIAL LOG

	Date	Materials	Man Hours
1.	_____	_____	_____
2.	_____	_____	_____
3.	_____	_____	_____
4.	_____	_____	_____
5.	_____	_____	_____
6.	_____	_____	_____

Obtain appropriate startup sheet and demonstrate normal system operation for your instructor.

REPLACE A DIRECT DRIVE BLOWER MOTOR

STUDY MATERIAL
Chapters 9 & 14

LABORATORY NOTES
Replacing an HVAC system component of any kind is a common procedure in the service industry. To avoid replacing parts unnecessarily some companies will require two people to test a part before condemning it. Most service technicians will verify the failure mode before replacing at the time of replacement. Sometimes this may seem a bit redundant but it may depend on the nature of the failure. Since motors are some of the more expensive replacement parts, it is probably a good idea to double-check the failure and demonstrate the failure mode to the customer as available. Some states require return of the replaced parts to the customer for their inspection. All of this is done to avoid replacing parts that do not need replacement. You may think this is no big deal but to some customers it is a very big deal. You should make it a practice to correctly identify a failure mode and not to replace parts at random.

UNIT DATA

1. Unit name _____   Model # _____   Serial # _____

2. Unit type _____

3. Part name _____   Model # _____   Serial # _____

VERIFY/DEMONSTRATE FAILURE MODE

Check	Step	Procedure
_____	1.	Describe precisely what the problem is. _____ _____
_____	2.	Make a distinction between the cause and the result. For example, an open winding would cause a motor to be supplied with voltage and not run; a seized up bearing would cause very high amps and a fuse failure; a dry bearing would cause the motor to turn slow and pull high amps.
_____	3.	If the problem is high motor amps, operate motor with all panels in normal position and measure amps.

Check	Step	Procedure
_____	4.	What test and result was used to verify the failure? _____
		_____
_____	5.	Certify that the part is damaged beyond repair and needs to be replaced.

REPLACE THE MOTOR

Check	Step	Procedure
_____	1.	Turn off power at accessible disconnect.
_____	2.	Lock out and tag at disconnect.
_____	3.	Measure voltage L1 and L2, L1 to ground, etc. Check that voltage is off.
_____	4.	Remove and carefully store all panels and covers.
_____	5.	Double check replacement part by manufacturer number, replacement part number, or physical and electrical characteristics.
_____	6.	Refer to wiring diagram to remove, unplug, or cut wires to the motor or blower assembly at the best location.
_____	7.	Unplug, remove, or cut wires as required.
_____	8.	Remove bolts or screws holding blower assembly.
_____	9.	Slide out blower assembly.
_____	10.	Remove screws or bolts holding motor within blower.
_____	11.	Loosen bolt holding blower wheel on motor.
_____	12.	Remove motor from blower assembly.
_____	13.	Install motor support bracket onto new motor.
_____	14.	Read and record motor full load amps (to be used later). Full load amps = _____
_____	15.	Start motor shaft into the blower wheel.
_____	16.	Position motor so that any oil holes will be accessible.
_____	17.	Line up bolt with flat side of shaft or insert key.
_____	18.	Slide motor and bracket into blower assembly.
_____	19.	Insert motor mounting screws or bolts.
_____	20.	Slide blower wheel into center and tighten bolt.
_____	21.	Spin blower by hand and inspect for wobble and correct position.
_____	22.	(Permanent split capacitor motors only) Install new capacitor with the new motor.
_____	23.	Slide the blower assembly back into equipment.
_____	24.	Plug in or reconnect wires to motor.
_____	25.	Double check color code to be sure motor is wired correctly.

MOTOR OPERATION CHECK

Check	Step	Procedure
_____	1.	Write with a permanent marker, new motor amperage in a service accessible location.
_____	2.	Turn on main power.
_____	3.	Install ammeter on L1 or wire going to terminal C of motor.
_____	4.	Obtain fan only operation.
_____	5.	Put all panels in normal position to obtain normal airflow.
_____	6.	Measure and record motor full load amps with all panels in position. Measured amps = _____
_____	7.	Compare measured amps with rated amps.

BELT DRIVE BLOWER SERVICE INSPECTION

STUDY MATERIAL
Chapters 9 & 14

LABORATORY NOTES
Some older residential and most commercial air handlers use belt drive blowers to deliver air through the duct system. The advantage of belt drive blowers over direct drive is the flexibility in choosing a blower speed and the ease of manufacturing a blower wheel to withstand the RPM. It is too difficult to match all the RPM, CFM, motor HP, etc., in all commercial applications. Most new residential systems have gone over to direct drive blowers but in commercial system belt drives will be around for many years to come. They require some different service procedures.

UNIT DATA

1. Unit name _____ Model # _____ Serial # _____

2. Unit type _____

3. Blower motor data: Voltage = _____ HP = _____ Locked rotor amps = _____

   Rated load amps = _____ Full load amps = _____

4. Blower wheel data: Width = _____ Diameter = _____ Shaft size OD = _____

5. Pulley size: Motor sheave OD = _____ Blower shaft sheave OD = _____

6. Belt size = _____ Number of belts = _____

ORIGINAL OPERATION/INSPECTION

Check	Step	Procedure
_____	1.	Spin blower wheel slowly by hand.
_____	2.	Note any drag noise or pulley wobble.
_____	3.	Install ammeter on L1 to motor.
_____	4.	Obtain fan only operation.
_____	5.	Observe motor begin to turn and watch for belt slippage during startup.

Check	Step	Procedure
_____	6.	Note any running noise, bearing, grinding, or any noise other than normal airflow noise.
_____	7.	Observe belt moving for any slippage, jumping, or abnormal movement.

DISASSEMBLY AND CLEANING

Check	Step	Procedure
_____	1.	Turn off power at accessible disconnect.
_____	2.	Lock out and install lockout tag.
_____	3.	Measure voltages L1 and L2, L1 to ground, etc. Make sure that voltage is off.
_____	4.	Remove and carefully store all panels and covers.
_____	5.	Loosen tension on belt for removal.
_____	6.	Remove belt and inspect for any cracks or severe shine (caused by slipping).
_____	7.	Inspect sheaves for any wear grooves.
_____	8.	Obtain and install replacement belts and sheaves with any sign of unusual wear.
_____	9.	Inspect blower shaft and bearings for signs of wear or noise. Replace as required.
_____	10.	Inspect blower wheel blades and scrape with a screwdriver or thin tool for any evidence of accumulated debris.
_____	11.	Clean blower wheel with water, high pressure air, or CO_2, whichever is most available and appropriate.
_____	12.	Blow motor air passages with compressed air or CO_2.
_____	13.	Wipe motor and nameplate clean with a rag (motor needs to be clean for cooling).
_____	14.	Install replacement sheaves as required.
_____	16.	Align or check alignment of pulleys with a straight edge on outer edge. The straight edge should touch on the outer edge of both pulleys, four places total.
_____	17.	Lube motor and shaft bearings as required.
_____	18.	Install new belt of correct size. Belt size = _____
_____	19.	Adjust motor tension assembly for correct tension. Belt play of 2–3 in is average. Consult manufacturer of blower for application and recommended tension.
_____	20.	Replace any loose or worn components of belt tension assembly as required.

CHECK FOR CORRECT OPERATION OF BLOWER AND MOTOR

Check	Step	Procedure
_____	1.	With a permanent marker, write the new motor amperage in a service accessible location.
_____	2.	Turn on main power.
_____	3.	Install ammeter on L1 or wire going to terminal C of motor.
_____	4.	Obtain fan only operation.

_____ 5.   Put all panels in normal position to obtain normal airflow.

_____ 6.   Measure and record motor full load amps with all panels in position.   Measured amps = _____

_____ 7.   Compare measured amps with rated amps.

INCREASE AIRFLOW BY MOTOR FULL LOAD AMPS (AS REQUIRED)

Frequently low airflow is the source of poor cooling or poor heat problems. When necessary the airflow of a belt drive blower can be increased by increasing the size of the motor sheave. Motor amperage must be checked with belt tension readjusted and all panels in their normal air delivery position. If more air is needed, replace the motor pulley and motor as required and within the limits of the blower wheel size.

Check	Step	Procedure
_____	1.	With a permanent marker, write the new motor amperage in a service accessible location.
_____	2.	Turn on main power.
_____	3.	Install ammeter on L1 or wire going to terminal C of motor.
_____	4.	Put all panels in normal position to obtain normal airflow.
_____	5.	Turn on blower.
_____	6.	Measure and record motor amps. Initial motor amps = _____
_____	7.	Turn off blower, then loosen and remove belt.
_____	8.	To get more air and raise amperage, adjust motor pulley to a larger size or obtain a larger motor pulley.
_____	9.	To reduce amperage and get a lower airflow decrease the pulley size.
_____	10.	Install belt and adjust tension. A belt too tight will make a rumbling noise and cause high motor amperage. A belt too loose will slip on startup and jump around on the slack side during operation.
_____	11.	Measure motor amps and compare with motor full load amps. Amps = _____
_____	12.	Repeat as required. Motor rated full load amps = _____ Final motor amps = _____

AIR CONDITIONING PERFORMANCE TEST

STUDY MATERIAL
Chapters 16 & 17

LABORATORY NOTES
The AC performance test is a capacity test of an operating AC unit. The test is begun by doing a startup and check operation on the system. The capacity of a system can be measured by airflow meters, ammeters, voltmeters, and using a pocket calculator. From our study of enthalpy, we use the formula NRE = HDC – HOC. The HDC (heat discharged from condenser) can be measured by CFM and temperature difference using the sensible heat formula. The HOC (heat of compression) can be measured by amps × volts × PF (power factor) of total system amp draw. The NRE or net refrigeration effect will be the actual capacity of the system.

UNIT DATA

1. Make _____ Model # _____ Serial # _____

2. Btu rating _____

STARTUP

Check	Step	Procedure
_____	1.	Install gauges on system.
_____	2.	Obtain power supply to system.
_____	3.	Inspect system for crankcase heater.
_____	4.	If crankcase heater is present, measure amps. Amps = _____
_____	5.	Make sure the crankcase heater has been energized for 8 hr minimum, or use the jog-start procedure. The jog-start procedure is as follows: start compressor by depressing contactor and run for 2 sec, turn off for 1 min, let oil settle, repeat process three times.
_____	6.	Obtain system normal cooling mode.
_____	7.	Allow system stabilization and record pressures.
		High side pressure = _____ Low side pressure = _____
_____	8.	Verify system normal operation.

MEASURE HEAT OF COMPRESSION (HOC)

Check	Step	Procedure
_____	1.	Measure amps:

        Compressor = _____

        Outside fan motor = _____

        Indoor fan motor = _____

        Total = _____

Check	Step	Procedure
_____	2.	Measure voltage at L1 and L2.
_____	3.	Estimate power factor (PF) at .8.
_____	4.	For single phase equipment use the following formula to determine the heat of compression. *Note that 3.14 is the conversion factor for getting our answer in the proper units, Btu.*

$$\text{Amps} \times \text{Volts} \times \text{PF} \times 3.14 = \text{HOC in Btu}$$

_____ × _____ × _____ × _____ = _____

Check	Step	Procedure
_____	5.	For three phase equipment use the following formula to determine the heat of compression:

$$\text{Amps} \times 1.83 \times \text{PF} \times 3.14 = \text{HOC in Btu}$$

_____ × _____ × _____ × _____ = _____

MEASURE HEAT DISCHARGED FROM CONDENSER (HDC)

Check	Step	Procedure
_____	1.	Measure airflow in CFM off condenser velocity × free area in ft².

_____ × _____ = CFM = _____

Check	Step	Procedure
_____	2.	Measure temperature difference across condenser.

        Air off condenser temperature – Ambient temperature = Temperature difference

_____ – _____ = _____

Check	Step	Procedure
_____	3.	CFM × 1.08 × Temperature difference = HDC in Btu

_____ × _____ × _____ = HDC = _____

CALCULATE NET REFRIGERATION EFFECT (NRE)

Check	Step	Procedure
_____	1.	NRE is the net refrigeration effect, or the cooling Btu.

NRE = HDC – HOC

_____ – _____ = NRE =_____

Check	Step	Procedure
_____	2.	Does NRE match with system rated capacity from # 2 from Unit Data above?

MEASURE NET REFRIGERATION EFFECT (NRE)

Check	Step	Procedure
_____	1.	Measure return air velocity and free area in ft^2.

Velocity × Area = Airflow (in CFM)

_____ × _____ = _____

Check	Step	Procedure
_____	2.	Measure supply air velocity free area in ft^2.

Velocity × Area = Airflow (in CFM)

_____ × _____ = _____

Check	Step	Procedure
_____	3.	The answers to steps 1 and 2 should be very close but probably not exactly the same.
_____	4.	Use average of 1 and 2 for total airflow. Airflow = _____ CFM
_____	5.	Calculate CFM per ton by the following formula:

$$\frac{\text{CFM from step 4 above}}{\text{Rated tons}} = \text{CFM per ton} = \underline{\hspace{1cm}}$$

Check	Step	Procedure
_____	6.	Measure and record the following temperatures: Return air wet bulb = _____ Return air dry bulb = _____ Supply air wet bulb = _____ Supply air dry bulb = _____
_____	7.	Using a psychrometric chart, plot return air condition point and supply air condition point.
_____	8.	Read and record enthalpy of return air (RAQ) and enthalpy of supply air (SAQ). RAQ = _____ SAQ = _____

Check	Step	Procedure
_____	9.	Use the total heat formula to calculate measured net refrigeration effect (NRE):

$4.5 \times$ CFM x (enthalpy of return air – enthalpy of supply air) = NRE

$4.5 \times$ CFM $\times$ (RAQ – SAQ)

$4.5 \times$ _____ $\times$ (_____ - _____) = NRE = _____ Btu

PLOT THEORETICAL APPARATUS DEW POINT

Check	Step	Procedure
_____	1.	Draw condition line through the return air and supply air condition points.
_____	2.	Continue condition line down to 100% saturation line of line of psychrometric chart.
_____	3.	This is your theoretical apparatus dew point (ADP). ADP = _____
_____	4.	How does this ADP compare with your operating low side pressure coil temperature?

_____

_____

PARTIAL REFRIGERANT RESTRICTION ON LIQUID LINE SIDE

STUDY MATERIAL
Chapter 9, Unit 2

LABORATORY NOTES

A partial restriction on the liquid line side of an AC system is a tricky thing to diagnose. The symptoms mimic the symptoms of low on refrigerant and many service technicians just add a little refrigerant, notice a little improvement, and call it good. Among experienced service people, it has always been considered a measure of a technician's level of skill to be able to diagnose and repair a partially restricted liquid line component.

The restriction generally occurs during a long term operation. The restriction material can be dirt, solder particles, rust flakes or any number of other solid particles. They will collect gradually. If the restriction were to happen all at once it would be felt in the system performance and someone would complain or notice. The restriction will cause liquid refrigerant to back up in the condenser or receiver if there is one. This refrigerant being taken out of circulation will cause the system to display symptoms of being low on charge: low side pressure low, high side pressure low, warm suction line, cool liquid line, and poor cooling.

The difference is that with a partial restriction in the liquid line, the refrigerant is not missing, it is simply not in circulation. Turn the system off, wait 5 min and turn it back on. The stored refrigerant will have leaked to the low side and now be pumped back to the high side. The restricted system will generally display a short term higher than normal high side pressure while the refrigerant is being collected in the liquid line. Other tip-offs will be high subcooling, any temperature drop across a restriction, and adding refrigerant that seems to disappear into the system.

The restriction occurs most commonly at a liquid line filter, capillary tube inlet screen, or TXV inlet screen, but could also occur at an elbow, soldered joint, or anywhere in a line. I once located a temperature drop in the middle of a liquid line and when I cut the line at that point, found a piece of rag that had been left in the line. The toughest to diagnose is the partially plugged inlet screen, because there is supposed to be a temperature drop at that point. The following series of tests and measurements will lead the service person to be able to better diagnose this not too common but frequently missed problem.

UNIT DATA

1. Unit make _____ Condenser model # _____ Coil model # _____

2. Unit type (circle one):   Split   Package   Window   Other _____

3. Refrigerant type: _____   Listed charge weight (if available): _____

4. Give the suction line size (OD) and the liquid line size (OD), and the length of the line:

   Suction line = _____ OD   Liquid line = _____ OD   Length = _____

5. Service complaint. _____

   _____

INITIAL SYSTEM SERVICE INSPECTION

Check	Step	Procedure
_____	1.	Obtain normal stabilized cooling mode.
_____	2.	Use the "determine normal operating pressures" section from Laboratory Worksheet AC-9 to determine normal expected operating pressures.
		High side pressure = _____ Low side pressure = _____
_____	3.	Measure and record the following. Measure liquid line temperature 6 in upstream of the liquid line filter and suction line temperature 12 in upstream from compressor.
		High side pressure = _____ Low side pressure = _____ Liquid line temperature = _____
		Suction temperature = _____

SUPERHEAT CALCULATIONS AND DIAGNOSIS

Check	Step	Procedure
_____	1.	Use a wet bulb thermometer to measure indoor wet bulb temperature.
		Wet bulb temperature = _____
_____	2.	Use a superheat calculator or chart to predict what the superheat should be. Superheat = _____
_____	3.	Using measured data from above, calculate the actual superheat by the following formula:

Suction line temperature – Coil temperature = Superheat

_____ – _____ = _____

Check	Step	Procedure
_____	4.	Compare expected superheat to actual measured superheat. The actual superheat is superheat is
		(circle one): Higher Same as Lower
		If higher or lower, by how much? _____°F

SUBCOOLING CALCULATIONS AND DIAGNOSIS

Check	Step	Procedure
_____	1.	Use temperature/pressure chart to determine the condensing temperature.
		High side condensing temperature = _____

Check	Step	Procedure
_____	2.	Calculate the maximum expected subcooling by using the following formula:

$$\text{Max subcooling} = \text{Ambient temperature} + \frac{\text{Ambient temperature} - \text{Condenser temperature}}{2} = \underline{\quad}$$

Check	Step	Procedure
_____	3.	Using measured data from step 3 above in the "Superheat Calculations and Diagnosis" section, calculate actual system operating subcooling. Subcooling = _____
_____	4.	Comparing maximum expected subcooling and actual measured subcooling, actual subcooling is (circle one): Higher Same as Lower If higher or lower, by how much? _____°F

TEMPERATURE DROP IN THE LIQUID LINE

Check	Step	Procedure
_____	1.	Verify continued normal system operation for at least 15 min.
_____	2.	Record system pressures again. High side pressure = _____ Low side pressure = _____
_____	3.	Measure and record liquid line temperature at the following locations: a. Condenser outlet temperature = _____ b. Liquid line filter inlet temperature = _____ c. Liquid line filter outlet temperature = _____
_____	4.	What is the temperature difference (in #3 above) from: a to c _____ b to c _____
_____	5.	Spot check before and after any fittings, soldered joints, and pinch points.
_____	6.	List and describe any temperature difference greater than 1°F. _____

TEST FOR ANY TEMPORARY RISE IN HIGH SIDE PRESSURE DURING STARTUP

Check	Step	Procedure
_____	1.	Verify continued normal system operation for at least 15 min.
_____	2.	Again record system pressures. High side pressure = _____ Low side pressure = _____
_____	3.	Turn off for 5 min.
_____	4.	Observe system pressures equalize.
_____	5.	Turn system on.
_____	6.	Does high side pressure go above high side pressure recorded in step # 2 above? (circle one): Yes No

REMOVE AND WEIGH CHARGE

This step is optional, to be used for extreme service problems.

Check	Step	Procedure
_____	1.	Use nameplate or manufacturer data to record exact weight of recommended charge.
		_____ lb _____.oz of R-_____
_____	2.	Obtain accurate scale and recovery equipment.
_____	3.	Recover and weigh existing charge: _____ lb _____ oz
_____	4.	Existing charge was (circle one):
		More than recommended Less than recommended The same as recommended

REPAIR AS REQUIRED

Check	Step	Procedure
_____	1.	If none of the tests performed above indicate partial restriction, proceed with a normal complete preventive maintenance service.
_____	2.	Cut out or remove and inspect any piece of liquid line, fitting, or capillary tube inlet screen that showed any temperature drop in liquid line.
_____	3.	Remove and inspect the inlet screen of the TXV if the system shows signs of restriction.
_____	4.	Remove TXV inlet screens or replace valve as required.
_____	5.	Install new liquid line filter if replacing TXV.
_____	6.	Replace liquid line filter if any temperature drop of 1°F or more was measured.
_____	7.	Clean condenser and evaporator coils as required.
_____	8.	Inspect system once more for any field alterations that could affect charge requirements or performance of system.

SYSTEM RECHARGE AND STARTUP

Check	Step	Procedure
_____	1.	Recharge system with virgin refrigerant. Weight = _____ lb _____ oz
_____	2.	Obtain normal cooling for a minimum of 10 min.
_____	3.	Measure and record final system operating conditions.
		Ambient temperature = _____
		High side pressure = _____
		Low side pressure = _____
		Liquid line temperature = _____
		Suction line temperature = _____
		Liquid line subcooling = _____

Suction line superheat = _____

Check	Step	Procedure
_____	4.	Compare current readings with original system service inspection from above.
_____	5.	Note any improvements or changes in system operation.
_____	6.	List all problems found and repairs made that address the original problem.

1. _____

2. _____

3. _____

4. _____

5. _____

FIELD TESTING PERMANENT SPLIT CAPACITOR AC COMPRESSORS

LABORATORY NOTES

Residential AC units, whether split, package, or self-contained, use permanent split capacitor motors because of the high motor efficiency. Both the permanent split capacitor motor and capacitor start/capacitor run motor use a run capacitor in series with the start winding at all times. Three phase compressors use no capacitors or relays at all. Pretty much all AC systems over $1\frac{1}{2}$ HP use a contactor to energize the compressor motor. In a typical service situation, a compressor can be, and frequently is, tested by manually engaging the contactor while measuring amperage of the compressor as it tries to start and run. Gauges should also be installed as the problem with a compressor can be with the pumping capacity and not the motor.

This laboratory worksheet leads the student through a field check out procedure as it would have to be done on the job site. This worksheet assumes that the service technician is on the job site and has diagnosed a possible problem in the compressor. In this situation, different from a bench compressor test, an applied voltage test would be performed first because the compressor already has wires and power connected to it. In the event of failing the applied voltage test, an ohms test will be performed to help diagnose the cause of failure.

APPLIED VOLTAGE TEST, PERMANENT SPLIT CAPACITOR

Check	Step	Procedure
_____	1.	Read and record from compressor nameplate:
		Rated load amps = _____ Locked rotor amps = _____
_____	2.	Using refrigerant type, ambient temperature, and manufacturer charts determine normal expected system operating pressures. High side pressure = _____ Low side pressure = _____
_____	3.	Install gauges on system. Record system idle pressure:
		High side pressure = _____ Low side pressure = _____
_____	4.	Remove wire from one side of the contactor coil.
_____	5.	Hand wire a toggle switch in series with the coil.
_____	6.	Main power to condensing unit or contactor is still off.
_____	7.	Turn thermostat to a call for cooling and obtain normal contactor coil voltage. Observe contactor snap in and out with toggle switch.
_____	8.	Turn toggle switch off so contactor is open.
_____	9.	Turn on main disconnect and measure voltage at L1 and L2.
_____	10.	Install ammeter on common lead of compressor.

Check	Step	Procedure
_____	11.	Turn toggle switch on and observe compressor try to start.
_____	12.	Observe compressor amperage and system pressures to determine compressor operation. Turn off within 5 sec if compressor fails to start and run.
_____	13.	Record system pressures and amperage if it does run. Compare with above.

High side pressure = _____   Low side pressure = _____   Amps = _____

Check	Step	Procedure
_____	14.	Use toggle switch to observe start and stop as required to help identify the problem.
_____	15.	If compressor fails to run or runs improperly, choose from the following conditions:

    _____ a. Pulls locked rotor amps and does not run. Diagnosis: stuck compressor.

    _____ b Runs but high amps. Diagnosis possibilities: partial short turn to turn, low supply voltage, check/replace run capacitor.

    _____ c. Does not run, pulls 0 amps. Diagnosis: open circuit (OL or winding).

    _____ d. Runs with low high side pressure and high low side pressure. Diagnosis: bad valves.

    _____ e. Runs with high high side pressure and high low side pressure. Diagnosis: overcharge.

    _____ f. Runs with low low side pressure and low high side pressure. Diagnosis: undercharge.

    _____ g. First e and quickly f. Diagnosis: partial blockage in liquid line or filter. Refer to Laboratory Worksheet AC-19.

In the event a compressor pulls LRA or does not run when supplied with voltage, an ohms test is in order. It is important to identify a cause of failure other than it won't run.

OHMS TEST, PERMANENT SPLIT CAPACITOR, CAPACITOR START/CAPACITOR RUN

Check	Step	Procedure
_____	1.	Turn off manual disconnect at AC unit.
_____	2.	Verify 0 voltage at L1 and L2 of contactor.
_____	3.	Remove compressor terminal cover and wires from all terminal connections.
_____	4.	Using the R × 10K ohms scale, measure and record ohms, terminal to frame:

1 = _____   2 = _____   3 = _____

Check	Step	Procedure
_____	5.	Determine start, run, and common compressor terminals.

Check	Step	Procedure
_____	6.	Record ohms test results and sketch terminal location in space below.

C = Terminal remaining while measuring highest ohms

S = Terminal of higher resistance from common terminal

R = Terminal of lowest resistance from common terminal

Ohm test 1 to 2 _____   C = _____

Ohm test 2 to 3 _____   R = _____

Ohm test 1 to 3 _____   S = _____

STARTING A STUCK COMPRESSOR

If you find a compressor that is stuck but checks out as OK on the ohms test, you can install a temporary or perhaps a permanent hard start kit. The hard start kit will use a start capacitor or an electronic device to give the compressor more start power during startup. It will be out of the circuit during normal run mode. Any permanent split capacitor motor type can have a hard kit installed but most do not need it. Pressures in a typical AC system equalize during the off cycle, but the system needs 2–3 min to do this. Any control problem or operator problem that calls for a short cycle restart can be a candidate for a permanent hard start kit.

Check	Step	Procedure
_____	1.	Rewire compressor to original wiring.
_____	2.	Obtain a hard start kit, preferably the one recommended by the compressor manufacturer. If no kit is available from the original manufacturer, generic solid state hard start kits are available.
_____	3.	Temporarily wire the start kit parallel to the run capacitor in the compressor circuit.
_____	4.	Install an ammeter on one lead to the hard start kit.
_____	5.	Use an unsharpened pencil with a full eraser to temporarily depress and close the manual contactor. *Note: Do not use any metal object to do this!*
_____	6.	Amperage through the start kit should occur for 1 sec.
_____	7.	If compressor starts and runs, let it run for about 1 min. Record amps: Full load amps = _____
_____	8.	If compressor does not run, check your wiring and obtain a different hard start kit. Repeat.
_____	9.	If it still won't run, condemn it.
_____	10.	If it does run, obtain a factory recommended hard start kit or install the field generic and check it out.
_____	11.	Some generic field replacement hard start kits require a cool down time and will not provide instant full power at restart. Be sure to get one that will start the compressor even when you are not there.

FIELD TESTING CAPACITOR START/CAPACITOR RUN AC COMPRESSORS

LABORATORY NOTES

A few residential and most light commercial AC units require a high starting torque compressor. Any unit that uses a TXV or for whatever reason needs an instant or short cycle full power restart must have a capacitor start/capacitor run or three phase motor. A capacitor start/capacitor run motor is a permanent split capacitor motor with the original run capacitor and a start capacitor wired to a potential relay, commonly called a hard start kit. Both motor types are powered by a contactor.

In a typical service situation, a compressor can be, and frequently is, tested by manually engaging the contactor while measuring amperage of the compressor as it tries to start and run. Gauges should also be installed as the problem with a compressor can be with the pumping capacity and not the motor.

This lab leads the student through a field check out procedure as it would have to be done on the job site. This job sheet assumes that the service technician is on the job site and has found a compressor that needs to be tested. In this situation an applied voltage test would be performed first because the compressor already has wires and power connected to it. In the event of failing the applied voltage test, an ohms test will be performed to help diagnose the cause of failure.

APPLIED VOLTAGE TEST, CAPACITOR START/CAPACITOR RUN

Check	Step	Procedure
_____	1.	Read and record from compressor nameplate:
		Rated load amps = _____ Locked rotor amps = _____
_____	2.	Using refrigerant type, ambient temperature, and manufacturer charts, determine normal expected
		system operating pressures. High side pressure = _____ Low side pressure = _____
_____	3.	Install gauges on system and record system idle pressure.
		High side pressure = _____ Low side pressure = _____
_____	4.	Remove wire from one side of the contactor coil.
_____	5.	Hand wire a toggle switch in series with the coil.
_____	6.	Main power to condensing unit or contactor is still off.
_____	7.	Turn thermostat to a call for cooling and obtain normal contactor coil voltage, and observe contactor snap in and out with toggle switch.
_____	8.	Turn toggle switch off so contactor is open.
_____	9.	Turn on main disconnect and measure voltage at L1 and L2: L1 = _____ L2 = _____

Check	Step	Procedure
_____	10.	Install ammeter on common lead of compressor.
_____	11.	Turn toggle switch on and observe compressor try to start.
_____	12.	Check compressor amperage and system pressures to determine compressor operation. Turn off within 3 sec if compressor fails to start and run.
_____	13.	Record system pressures and amperage if it does run. Compare with above.

High side pressure = _____   Low side pressure = _____   Amps = _____

_____	14.	If does not start, install ammeter on the single wire going from terminal 2 of the start relay to the start capacitor or a single wire from the start capacitor to the run capacitor.
_____	15.	Attempt to start compressor. Observe amperage through the start capacitor. It should go to 2–6 amps and drop to 0 within 2 sec.
_____	16.	If amps stays high, relay is not disengaging.
_____	17.	If amps stay at 0, relay is not engaging.
_____	18.	Equalize system pressures through the gauge manifold and energize compressor.
_____	19.	If it runs, record amps at common lead. Amps = _____
_____	20.	Disconnect existing start capacitor and relay.
_____	21.	Obtain a 230 V or higher, solid state hard start kit. Look for a Supco Super Boost, SPP-7S or SPP8 or equivalent.
_____	22.	Temporarily wire the start kit parallel to the run capacitor in the compressor circuit.
_____	23.	Install an ammeter on one lead to the hard start kit.
_____	24.	If compressor runs and starts under load using the generic replacement hard start kit, permanently replace the original with the generic or obtain an original replacement. Generally speaking, the original will give better performance than any generic.
_____	25.	If compressor fails to run or runs improperly, choose from the following possible diagnoses:

_____   a. Pulls locked rotor amps and does not run. Diagnosis: stuck compressor.

_____   b. Runs but high amps. Diagnosis: partial short turn to turn.

_____   c. Does not run, pulls 0 amps. Diagnosis: open (OL or winding).

_____   d. Runs with low high side pressure and high low side pressure. Diagnosis: bad valves.

_____   e. Runs with high high side pressure and high low side pressure. Diagnosis: overcharge.

_____   f. Runs with low low side pressure and low high side pressure. Diagnosis: undercharge.

_____   g. The high side pressure goes up and then both high side pressure and low side pressure go to lower than normal. Diagnosis: Suspect a partial blockage and perform Laboratory Worksheet AC-19 lab on system.

OHMS TEST, PSC/CSR (AS REQUIRED)

Check	Step	Procedure
_____	1.	If compressor fails to start, perform an ohms test.
_____	2.	Turn off manual disconnect at AC unit.
_____	3.	Verify 0 V at L1 and L2 of contactor.
_____	4.	Remove compressor terminal cover and wires from all terminal connections.
_____	5.	Using the $R \times 10K$ ohms scale, measure and record ohms, terminal to frame.
		1 = _____ 2 = _____ 3 = _____
_____	6.	Determine start, run, and common compressor terminals.
_____	7.	Record ohm test results and sketch terminal location in space below.

C = Terminal remaining while measuring highest ohms

S = Terminal of higher resistance from common terminal

R = Terminal of lowest resistance from common terminal

Ohm test 1 to 2 _____   C = _____

Ohm test 2 to 3 _____   R = _____

Ohm test 1 to 3 _____   S = _____

FIELD TESTING THREE PHASE AC COMPRESSORS

LABORATORY NOTES

Many commercial AC systems and a few residential types use three phase compressor motors. They are stronger and simpler to use in that they do not require any extra hard start or efficiency components. Three phase motors have good running efficiency and plenty of start power without the addition of these devices. The most common failure mode of three phase compressors is loss of one leg, called single phasing. Compressor testing is done simply to verify that three phase power is available and to be sure it gets to the compressor.

APPLIED VOLTAGE TEST, THREE PHASE

Check	Step	Procedure
_____	1.	Read and record data from the compressor nameplate.
		Rated load amps = _____
		Locked rotor amps = _____
		Voltage = _____
		Phase (circle one): Single phase Three phase
_____	2.	Using refrigerant type, ambient temperature, and manufacturer charts, determine normal expected system operating pressures. High side pressure = _____ Low side pressure = _____
_____	3.	Install gauges on system and record system idle pressure.
		High side pressure = _____ Low side pressure = _____
_____	4.	Remove wire from one side of the contactor coil.
_____	5.	Hand wire a toggle switch in series with the coil.
_____	4.	Main power to condensing unit or contactor is still off.
_____	5.	Turn thermostat to a call for cooling and obtain normal contactor coil voltage. Observe contactor snap in and out with toggle switch.
_____	6.	Turn toggle switch off so contactor is open.
_____	7.	Turn on main disconnect and measure voltage at the following locations:
		L1 to L2 = _____ L2 to L3 = _____ L1 to L3 = _____
_____	8.	Install ammeter on any one lead of compressor.
_____	9.	Turn toggle switch and observe compressor try to start.

Check	Step	Procedure
_____	10.	Observe condition (circle one):

Compressor will start and run   Tries to start and blows fuse   Nothing happens

Check	Step	Procedure
_____	11.	If compressor starts and runs, measure and record the following amp and pressure information:

L1 amps = _____   L2 amps = _____   L3 amps = _____   High side pressure = _____

Low side pressure = _____

Check	Step	Procedure
_____	12.	If compressor fails to run or runs improperly, choose one of the following diagnoses:

_____   a. Pulls locked rotor amps and does not run. Diagnosis: stuck compressor.

_____   b. Runs but high amps. Diagnosis: partial short turn to turn.

_____   c. Does not run, pulls 0 amps. Diagnosis: open (OL or winding).

_____   d. Runs with low high side pressure and high low side pressure. Diagnosis: bad valves.

_____   e. Runs with high high side pressure and high low side pressure. Diagnosis: overcharge.

_____   f. Runs with low low side pressure and low high side pressure. Diagnosis: undercharge.

_____   g. First e and quickly f. Diagnosis: partial blockage in liquid line or filter. Refer to Laboratory Worksheet AC-19.

OHMS TEST, THREE PHASE (FOR COMPRESSORS THAT DO NOT RUN)

Check	Step	Procedure
_____	1.	Turn off manual disconnect at AC unit.
_____	2.	Verify 0 voltage at L1/L2, L2/L3, and L1/L3.
_____	3.	Remove compressor terminal cover and wires from all terminal connections.
_____	4.	Using the R × 10K ohms scale, measure and record ohms, terminal to frame.

1 = _____   2 = _____   3 = _____

Check	Step	Procedure
_____	5.	Measure and record ohms of all three possibilities of terminal measurement terminals.

1–2 = _____   2–3 = _____   1–3 = _____

Check	Step	Procedure
_____	6.	All three ohms measurements from step #5 should be identical and more than zero.

CONTROLS LABORATORIES

BASIC METER USE

STUDY MATERIAL
Chapter 2, Unit 2

LABORATORY NOTES
See your instructor for a box of parts to be tested and standard testing stations set up at various shop equipment locations.

VOLTMETER 120/240
Use a voltmeter to measure voltage at the requested locations. Set the meter at the highest scale until the approximate voltage is known to avoid pegging the meter.

120 V duplex right to left = _____

120 V duplex right to ground = _____

120 V duplex left to ground = _____

240 V single phase right to left = _____

240 V single phase right to ground = _____

240 V single phase left to ground = _____

VOLTMETER 240 V THREE PHASE
Test the voltage at the following points:

L1 to L2 = _____   L1 to ground = _____

L1 to L3 = _____   L2 to ground = _____

L2 to L3 = _____   L3 to ground = _____

VOLTMETER 480 V THREE PHASE
Test the voltage at the following points:

L1 to L2 = _____   L1 to ground = _____

L1 to L3 = _____   L2 to ground = _____

L2 to L3 = _____   L3 to ground = _____

OHMMETER (TEST AT OPEN TABLE)
Use an ohmmeter to measure the resistance of the listed components and locations. A value of 0 ohms must test 0 on the R1 scale with a meter zeroed on the R1 scale. Anything over 10,000 ohms is high resistance. Obtain exact ohm values for all others.

Potential (contacts) 1 to 2 = _____

Relay (coil) 2 to 5 = _____

Amperage relay coil = _____

Contacts = _____

Transformer line side = _____

Load side = _____

Line to load = _____

Compressor line voltage overload relay

C to R = _____

Overload contacts OK? (circle one):   Yes   No

R to S = _____

C to S = _____

Main power = _____

C to dome = _____

R to dome = _____

S to dome = _____

MILLIVOLT METER
Perform this test at a burner simulator. Use a millivolt meter to perform an open circuit test on a thermocouple. Measure the voltage from the core to the shell while heating the thermocouple with a pilot flame or torch.

Voltage = _____ mV

THERMOSTAT HEAT CIRCUIT AMPERAGE

Perform this test at any operable furnace.

Check	Step	Procedure
_____	1.	Use a hook jaw analog ammeter and multiple coils of wire connected from R to W of thermostat subbase to measure thermostat amperage. Amp reading = _____ Number of Wraps = _____ Thermostat amps = _____
_____	2.	Use a digital in line ammeter to measure stat amps. Test one loop of jumper wire R to W. Amps = _____

AMPS AND VOLTS OF A DIRECT DRIVE FURNACE BLOWER

Use an ammeter to measure the amperage of a fan motor. The fan motor must be operating to pull amperage. Close off airflow with a damper or panel and observe the amperage change. Does the current increase or decrease as dampers are closed off? Measure amps on white wire (neutral). Use the voltmeter to measure voltage from the unused speed tap of the motor to ground.

Normal amps = _____

Closed damper amps = _____

Increase/decrease = _____

Neutral amps = _____

Speed tap voltage = _____

BASIC VOLT OHM METER USE

STUDY MATERIAL
Chapter 7, Unit 1

LABORATORY NOTES

This laboratory worksheet will cover basic volt ohm meter (VOM) use. Use your own VOM if you have one. This will be practice in using it. Repeat the lab with other meters if you will be using several different VOMs in the length of the course.

Check	Step	Procedure
_____	1.	Obtain VOM, Honeywell 240/208/120 V x 24 V transformer (Honeywell AT140 1018), one R8222D 1014 general purpose relay, one toggle switch, and assorted power supply plugs and wires.
_____	2.	List meter name and number. _____
_____	3.	List all scales. _____
_____	4.	Test and list required batteries. _____

TEST OHMMETER SCALES (SET METER TO APPROPRIATE OHMS SCALE)

Check	Step	Procedure
_____	1.	Measure and record ohms of the following circuits and switches:

    a. 240 V (orange) to C (black) on primary of transformer.   Ohms = _____

    b. 208 V (red) to C (black) on primary of transformer.   Ohms = _____

    c. 120 V (white) to C (black) on primary of transformer.   Ohms = _____

    d. R and C on secondary or load side of transformer.   Ohms = _____

    e. Coil terminals of R8222D1014 relay.   Ohms = _____

    f. Terminals 1 and 3 NO contacts R8222D1014 relay.   Ohms = _____

		g. Terminals <u>4</u> and <u>6</u> NO contacts R8222D1014 relay.	Ohms = _____
		h. Terminals <u>1</u> and <u>2</u> NC contacts R8222D1014 relay.	Ohms = _____
		i. Terminals <u>4</u> and <u>5</u> NC contacts R8222D1014 relay.	Ohms = _____

MEASURE VOLTAGE

For the following procedure, set the meter to the appropriate voltage setting.

Check	Step	Procedure

_____ 1. Measure and record voltage at the following terminal locations.

a. Measure voltage from a 120 V receptacle. Voltage = _____

b. Connect 120 V to orange and black wires of transformer. Read voltage at line and load terminals of transformer.   Line voltage = _____   Load voltage = _____

c. Connect 12 V to white and black wires of transformer. Read voltage at line and load terminals of transformer.   Line voltage = _____   Load voltage = _____

d. Connect 240 or 208 V to appropriate terminals of transformer to obtain 24 V at load terminals of transformer.   Line voltage = _____   Load voltage = _____

e. Remove power from transformer before continuing to step f.

f. Using a red thermostat wire, connect R of transformer to power side of toggle.

g. Using a blue thermostat wire, connect switch side of toggle to coil of relay.

h. Using a yellow thermostat wire, connect the other coil terminal to C of the transformer.

i. With toggle turned off, supply 120 V to line side of transformer, white and black.

j. Measure 0 V at coil of relay.

k. Turn VOM to ohm scale. Measure infinite ohms at NO contacts, terminals 4 and 6.

l. Measure 0 ohms at NC contacts, terminals <u>4</u> and <u>5</u>.

m. Turn VOM to voltage scale. While measuring voltage at coil of relay turn toggle switch to on. Measure 24 V at coil.

n. Turn VOM to ohm scale. Measure 0 ohms at NO contacts, terminals 4 and 6.

o. Measure infinite ohms at NC contacts, terminals <u>4</u> and <u>5</u>.

p. Turn the toggle off and back. You should hear the relay clicking as the contacts change position.

q. With the toggle turned off, measure the following voltages:

- Voltage supplied to circuit. C to R at transformer. Voltage should be 24 V. Voltage = _____
- Voltage supplied to load. Coil terminals at relay. Voltage should be 0 V.   Voltage = _____
- Trace circuit using C as L2. C to red at switch. Voltage should be 24 V.   Voltage = _____
- Continue tracing circuit. C to blue at switch. Voltage should equal 0 V.   Voltage = _____

This process of troubleshooting with a voltmeter is called hopscotching and is the preferred method of locating an open switch used by many service technicians. The voltage of 0 at C to red switch followed by 24 V at C to blue switch proves that the switch has voltage going to it but not coming out. This proves that the switch is open and when closed will pass the voltage on down the circuit. It is possible that another switch may be open and the load not be energized. In this circuit the single toggle is the only switch in series with the load and turning the switch on and off will energize the relay coil. Later, when this method is applied to other circuits, continued troubleshooting will be needed.

Question: What is the problem if 0 volts is read from C to red switch and 0 volts is read from C to blue switch?

Answer: Voltage is not making it to switch, repeat test at C and R of power supply.

TESTING CONTINUITY OF ELECTRICAL COMPONENTS

STUDY MATERIAL
Chapter 7, Unit 1

LABORATORY NOTES
Use an ohmmeter to test the various circuits of electrical components used in HVAC systems. Make requested tests and record exact numbers of ohms from the requested circuit. From the results of these tests you will make conclusions about the quality of the part. Use exact terms such as shorted coil, open contacts, grounded to case, welded contacts, etc. Use OK or GOOD to describe a part that checks out on the continuity test. Keep in mind that an applied voltage test of a component may show other defects. Materials needed: ohmmeter, assorted electrical parts, and ohmmeter.

Test first line break overload

1. Make _____   Part # _____

2. Test 1 to 2.   Ohms = _____

3. Test 1 to 3.   Ohms = _____

   Conclusion: _____

   _____

Test second line break overload

1. Make _____   Part # _____

2. Test 1 to 2.   Ohms = _____

3. Test 1 to 3.   Ohms = _____

   Conclusion: _____

   _____

Test first potential relay

1. Make _____   Part # _____

2. Test 1 to 2.   Ohms = _____

3. Test 2 to 5.   Ohms = _____

   Conclusion: _____

   _____

Test second potential relay

1. Make _____   Part # _____

2. Test 1 to 2.   Ohms = _____

3. Test 2 to 5.   Ohms = _____

   Conclusion: _____

   _____

Test first amperage magnetic relay

1. Make _____   Part # _____

2. Test L-M (coil).   Ohms = _____

3. Test S to L (contacts).   Ohms = _____

   Conclusion: _____

   _____

Test second relay

1. Make _____   Part # _____

2. Test L-M (coil).   Ohms = _____

3. Test S to L (contacts).   Ohms = _____

   Conclusion: _____

   _____

Test first transformer

1. Make _____   Part # _____

2. Primary voltage = _____   Secondary voltage = _____   VA = _____

3. Test primary coil.   Ohms = _____

4. Secondary coil.   Ohms = _____

5. Test terminal to case.   Ohms = _____

6. Test primary to secondary. Ohms = _____

   Conclusion: _____

   _____

Test second transformer

1. Make _____   Part # _____
2. Primary voltage = _____   Secondary voltage = _____   VA = _____
3. Test primary coil.   Ohms = _____
4. Test secondary coil.   Ohms = _____
5. Test terminal to case.   Ohms = _____
6. Test primary to secondary.   Ohms = _____

   Conclusion: _____

   _____

Test first overload relay

1. Make _____   Part # _____
2. Test 1–2 (line).   Ohms = _____
3. Test 3–4 (OL contacts).   Ohms = _____

   Conclusion: _____

   _____

Test second overload relay

1. Make _____   Part # _____
2. Test 1–2 (line).   Ohms = _____
3. Test 3–4 (OL contacts).   Ohms = _____

   Conclusion: _____

   _____

Test first magnetic starter

1. Make _____   Part # _____   Heater # _____
2. Test coil voltage.   Coil ohms = _____   Number of poles = _____
3. Manually depress contactor portion and test ohms of the following terminals:

   L1–T1 ohms = _____   L2–T2 ohms = _____   L3–T3 ohms = _____
4. Test overload relay load circuit   L1 ohms = _____   L2 ohms = _____   L3 ohms = _____
5. Test overload relay control circuit.   L1 ohms = _____   L2 ohms = _____   L3 ohms = _____

Test second magnetic starter

1. Make _____   Part # _____   Heater # _____
2. Test coil voltage.   Coil ohms = _____   Number of poles = _____
3. Manually depress contactor portion and test ohms of the following terminals:

   L1–T1 ohms = _____   L2–T2 ohms = _____   L3–T3 ohms = _____
4. Test overload relay load circuit.   L1 ohms = _____   L2 ohms = _____   L3 ohms = _____
5. Test overload relay control circuit.   L1 ohms = _____   L2 ohms = _____   L3 ohms = _____

Test first single phase compressor

_____ 1.   Draw a sketch of the terminals.

_____ 2.   Label the terminals 1, 2, and 3.

_____ 3.   Measure and record ohms for each terminal pair.

                1–2 ohms = _____   2–3 ohms = _____   1–3 ohms = _____

_____ 4.   The run winding is the lowest resistance. (Circle the run winding):   1–2   2–3   1–3

_____ 5.   The start winding is the middle resistance. (Circle the start winding): 1–2   2–3   1–3

_____ 6.   The run winding resistance added to the start winding resistance should be equal to the resistance of the third terminal pair. Is it? (circle one):   Yes   No

_____ 7.   The two terminals reading the highest resistance must be the single end of run and start. The terminal left over is the common terminal.   C = _____

_____ 8.   C to the higher resistance is the start terminal.   S = _____

_____ 9.   C to the lowest resistance is the run terminal.   R = _____

Test a second single phase compressor

_____ 1.   Draw a sketch of the terminals.

_____ 2.   Label the terminals 1, 2, and 3.

_____ 3.   Measure and record ohms for each terminal pair.

                1–2 ohms = _____   2–3 ohms = _____   1–3 ohms = _____

_____ 4.   The run winding is the lowest resistance. (Circle the run winding):   1–2   2–3   1–3

_____ 5.   The start winding is the middle resistance. (Circle the start winding): 1–2   2–3   1–3

_____ 6.   The run winding resistance added to the start winding resistance should be equal to the resistance of the third terminal pair. Is it? (circle one):   Yes   No

_____ 7.   The two terminals reading the highest resistance must be the single end of run and start. The terminal left over is the common terminal.   C = _____

_____ 8.   C to the higher resistance is the start terminal.   S = _____

_____ 9.   C to the lowest resistance is the run terminal.   R = _____

TESTING CAPACITORS, FOUR METHODS

STUDY MATERIAL
Chapter 7, Unit 4

LABORATORY NOTES
This laboratory worksheet will cover four methods of measuring capacitance, the ohmmeter test; the applied voltage test measured two ways, with a spark and by measuring voltage; and the capacitor test. The laboratory worksheet will go on to cover capacitors in parallel, capacitors in series, and circuits with a combination of series and parallel capacitors.

Part 1: OHMMETER TEST
The first capacitor test method involves testing with an analog ohmmeter. The capacitor must be first shorted out by crossing the two terminals with a screwdriver or any other metal tool with a rubber or plastic handle for insulation. The capacitor is tested by placing the two test leads across the two capacitor terminals. A good capacitor will show a deflection of the needle from 0 ohms toward infinite and then back toward 0. Start with the 0 to 1K but try several ohms scales to see the deflection. If you get a deflection on any scale, you have a good capacitor. This method is the traditional HVAC service technician method and is quite accurate. Discard any capacitors showing signs of physical damage; splits, bulges, cracks, leaking, or loose or broken terminals, etc. In the service world about 75% of bad capacitors can be condemned by a close visual inspection.

Check	Step	Procedure
_____	1.	Obtain an analog ohmmeter or multimeter with several ohms scales and an assortment of capacitors.
_____	2.	Discharge each capacitor after each test.
_____	3.	Record capacitor data as requested.
_____	4.	Test each capacitor on various scales.
_____	5.	A good capacitor will show a steady movement in the meter, first toward 0 and then, as the capacitor charges, toward infinity.

Check	Step	Procedure

6. Test five assorted start capacitors. List the capacitance in microfarads, the voltage, the length and diameter of the capacitor, the deflection in ohms, and the problem with the capacitor, if any.

	MFD	V$_{AC}$	Length	Diameter	Deflection	Problem (if any)
a.	—	—	——	——	——	——
b.	—	—	——	——	——	——
c.	—	—	——	——	——	——
d.	—	—	——	——	——	——
e.	—	—	——	——	——	——

7. What relationship does physical size have on the capacitance value of start capacitors? Does a big capacitor mean a large capacitance in microfarads? (circle one):   Yes   No

8. What relationship does physical size have on the V$_{AC}$ rating? Does a big capacitor mean a big voltage? (circle one):   Yes   No

9. What relationship does physical size have on distance of deflection? Does a high capacitance in microfarads mean a greater distance of deflection? (circle one):   Yes   No

10. Test five assorted run capacitors. List the capacitance in microfarads, the voltage, the length and diameter of the capacitor, the deflection in ohms, and the problem with the capacitor, if any.

	MFD	V$_{AC}$	Length	Diameter	Deflection	Problem (if any)
a.	—	—	——	——	——	——
b.	—	—	——	——	——	——
c.	—	—	——	——	——	——
d.	—	—	——	——	——	——
e.	—	—	——	——	——	——

11. What relationship does physical size have on the capacitance value of run capacitors? Does a big capacitor mean a large capacitance in microfarads? (circle one):   Yes   No

12. What relationship does physical size have on the V$_{AC}$ rating? Does a big capacitor mean a big voltage? (circle one):   Yes   No

13. What relationship does physical size have on distance of deflection? Does a high capacitance in microfarads mean a greater distance of deflection? (circle one):   Yes   No

Part 2: APPLIED VOLTAGE SPARK TEST (OPTIONAL)

The second method of testing capacitors is the hold a charge method or spark test. This is also an old method of testing capacitors that is actually quite accurate and primarily used by electricians. It is a simple test and requires only a line voltage test cord and a screwdriver. The test is performed by applying a charge to the capacitor, wait 5 min, and short it out. If it sparks it held the charge and is a good capacitor. Care must be taken when discharging the capacitor. The higher the microfarad rating, the bigger the spark.

Check	Step	Procedure
_____	1.	Select four capacitors to be tested with varying microfarad values.
_____	2.	Obtain a 120 V plug with two alligator clips or quick connect terminals on the L1 and L2 wires.
_____	3.	Capacitors are charged by applying a voltage (120 V) with a cord plug for about 1 sec only. Connect the capacitor to be tested, plug in, pull plug.
_____	4.	Allow 1–5 min of wait time and cross terminals.
_____	5.	A good capacitor will spark, a bad capacitor will not.
_____	6.	Write OK or Fail for each capacitor you test.
		Capacitor 1 _____ Capacitor 2 _____ Capacitor 3 _____ Capacitor 4 _____

Part 3: APPLIED VOLTAGE MEASURE AMPERAGE TEST

The third method of testing is the traditional preferred method of seasoned HVAC service technicians. It involves applying a voltage to the capacitor in exactly the same manner as the spark test but measuring the amperage flowing through the capacitor. This amperage and the applied voltage is then inserted into the following formula:

$$\text{Capacitance (in microfarads)} = \frac{2650 \times \text{Amps}}{\text{Applied voltage}}$$

Be sure to do the multiplication first and then the division. A calculator is recommended to simplify the math. This method was the way to find a 25 microfarad capacitor that was performing at 15 microfarads and working but affecting motor performance.

Check	Step	Procedure
_____	1.	Select two run and two start capacitors rated at 220 V_{AC} or higher.
_____	2.	Use capacitors that have passed both the ohms test and the spark test from above.
_____	3.	Some instructors/service technicians will put a complete wrap of electrical tape around a start capacitor before testing in this manner. Do not test the capacitor longer than 5 sec and do not retest if it is warm.
_____	4.	Perform the above applied voltage test and measure the amp draw through the capacitor.
_____	5.	Use the following formula for single phase:

$$\frac{2650 \times \text{amp}}{\text{volt}} = \text{MFD}$$

Check	Step	Procedure
_____	6.	Use the following formula for three phase:

$$\frac{3180 \times amp}{volt} = MFD$$

Check	Step	Procedure
_____	7.	Test four assorted capacitors. Applied voltage = _____

	Run/Start	Rated MFD	Amp	Volt	Performing MFD
a.	_____	_____	____	____	_____
b.	_____	_____	____	____	_____
c.	_____	_____	____	____	_____
d.	_____	_____	____	____	_____

Part 4: TESTING CAPACITORS WITH DIGITAL READOUT MICROFARAD TEST SCALE

We now have available testers and test scales that will give a digital readout of the performing capacitance in microfarads by touching two leads to the two terminals. That is the last test method in this sequence of four test methods. Modern digital multimeters are frequently equipped with a microfarad scale simplifying the capacitor test procedure. Applying the two test leads to an uncharged capacitor will cause the meter to read microfarads. A charged capacitor could damage the meter just as it could any ohmmeter or ohms scale of a meter.

Check	Step	Procedure
_____	1.	Use the same four capacitors as tested in step #7 of the applied voltage measure amperage test.
_____	2.	Obtain a digital multimeter with a microfarad scale.
_____	3.	Discharge and test the four capacitors.
_____	4.	Compare the digital meter readings with both the rated capacitance and the results of the previous capacitance test.

	Start/run	Rated MFD	Step #7 MFD	Digital MFD
a.	_____	_____	_____	_____
b.	_____	_____	_____	_____
c.	_____	_____	_____	_____
d.	_____	_____	_____	_____

Part 5: MEASURE CAPACITANCE OF TWO CAPACITORS IN PARALLEL

The next three tests do not involve a different method of testing but the test is performed on a bank of capacitors instead of a single capacitor. A bank of wired capacitors will perform as a single capacitor. Sometimes a service technician can assemble or build a capacitor of a required MFD rating that is not on the truck or available. Using multiple capacitors in a circuit to replace a single blown capacitor may cause a space or mounting problem. Some technicians would prefer to return later with the correct single capacitor to replace a built up bank of capacitors. For the test on this laboratory worksheet we will use only run capacitors. The same procedure could be applied to start capacitors.

Check	Step	Procedure
_____	1.	Select any two run capacitors.
_____	2.	Use preconstructed wires with quick connect terminals to wire two capacitors in parallel.
_____	3.	Use the following formula for two capacitors in parallel to calculated rated capacitance:

$$C1 + C2 = \text{Total parallel capacitance} = \_\_\_\_\_$$

Check	Step	Procedure
_____	4.	Using a digital tester measure the capacitance of the capacitor bank. Capacitance = _____ mfd
_____	5.	Use the following formula to measure the percent variation in measured capacitance compared to the rated capacitance:

$$\frac{\text{Measured capacitance}}{\text{Rated capacitance}} \times 100 = \text{Variation of } \_\_\_\_\_ \%$$

Check	Step	Procedure
_____	6.	How close does the applied voltage capacitance match up with the rated capacitance?

Part 6: MEASURE CAPACITANCE OF TWO CAPACITORS IN SERIES

Check	Step	Procedure
_____	1.	Select and wire any two run capacitors in series.
_____	2.	Use the formula for two capacitors in series to calculate the rated capacitance:

$$\frac{C1 \times C2}{C1 + C2} = \text{Total series capacitance} = \_\_\_\_\_$$

Check	Step	Procedure
_____	3.	Using a digital tester measure the capacitance of the series capacitor bank. Capacitance = _____ mfd
_____	4.	Use the following formula to calculate the percent of variation from the calculated capacitance:

$$\frac{\text{Measured capacitance}}{\text{Rated capacitance}} \times 100 = \text{Variation of } \_\_\_\_\_ \%$$

Check	Step	Procedure
_____	5.	How close does the applied voltage capacitance match up with the rated capacitance?

Part 7: MEASURE CAPACITANCE OF THREE CAPACITORS IN A SERIES/PARALLEL CIRCUIT

Check	Step	Procedure
_____	1.	Select and wire any three run capacitors in series/parallel circuit.
_____	2.	List the rated capacitance of capacitors used taking care to record C1 as C1, etc.

Check	Step	Procedure

_____ 3.   Use the formula for a series/parallel circuit to calculate the rated capacitance:

$$\text{Total series/ parallel capacitance} = \frac{C1 \times C2}{C1 + C2} + C3 = \text{\_\_\_\_\_}$$

_____ 4.   Using a digital tester measure the capacitance of the capacitor bank at the identified terminals.

Capacitance = _____ mfd

_____ 5.   Use the following formula to calculate the percent of variation from the calculate capacitance:

$$\frac{\text{Measured MFD}}{\text{Rated MFD}} \times 100 = \text{Variation of \_\_\_\_\_\%}$$

_____ 6.   How close does the applied voltage capacitance match up with the rated capacitance? _____

BLOWER SPEED COMPARISION

STUDY MATERIAL
Chapter 7, Units 3 & 4

LABORATORY NOTES

This laboratory worksheet will cover the comparison of blower speed. This worksheet covers making adjustments to belt drive blowers, changing pulleys, replacing shafts, and measuring amperage of the motor. The student will be asked to make comparisons and draw conclusions about how these adjustments affect the airflow and motor amp load. After completing this laboratory worksheet the student will be able to perform these blower service operations and correct airflow problems, originating in the blower, within the limits of the equipment. The blower used is a typical residential size blower, but the conclusions made hold true for any belt drive blower using a squirrel cage type blower wheel with forward curved blades.

   Material needed: One bench mounted blower assembly. Two blower motors (one permanent split capacitor and one split phase). Assorted belts, pulleys, and shafts. Allen wrenches in sizes of $1/8$ and $5/32$ in. Miscellaneous hand tools. Ammeter and tachometer.

MOTOR DATA

Split phase motor:

1.   HP _____   RPM _____

2.   Full load amps = _____   Locked rotor amps = _____

3.   Motor shaft size = _____

Permanent split capacitor motor:

1.   HP _____   RPM _____

2.   Full load amps = _____   Locked rotor amps = _____

3.   Motor shaft size = _____

Check	Step	Procedure
_____	1.	Install the split phase motor on the blower assembly. *Note: Do not install fan belt.*
_____	2.	Run the motor without the belt and record the following motor information.

Motor RPM = _____   Motor amps = _____

_____ 3. Install and adjust a motor pulley for a small diameter. Pulley sheaves should be far apart.

_____ 4. Install a belt for proper tension, just tight enough to avoid slipping during startup. Run motor and record the following information:

Blower RPM = _____   Motor RPM = _____   Motor amps = _____

_____ 5. Adjust the motor pulley in one full turn to make a larger pulley. This will cause a faster blower wheel speed. Run the motor and record the following information:

Blower RPM = _____   Motor RPM = _____   Motor amps = _____

_____ 6. Observe that as the motor load increases, the speed of the motor drops, while the amps of the motor increases.

_____ 7. Increase the size of the motor pulley to load the motor to its full load amp rating. Run the motor and record the following information. *(Note: Do not exceed motor rated full load amps).*

Blower RPM = _____   Motor RPM = _____   Motor amps = _____

_____ 8. Block off the discharge air of the blower with a piece of sheet metal and record the following

information.   Blower RPM = _____   Motor RPM = _____   Motor AMP = _____

_____ 9. What happens to the motor amps as the discharge air is blocked? _____

_____

_____ 10. Remove the discharge block and install the inlet block around the blower air intake.

_____ 11. Make a comparison of a blocked discharge (all dampers closed) and a blocked inlet (plugged filters).

_____ 12. Install the permanent split capacitor motor on the blower assembly and repeat steps 1 through 7 above.

	Blower RPM	Motor RPM	Motor amps
Speed 1 (no belt)	_____	_____	_____
Speed 2 (small pulley)	_____	_____	_____
Speed 3 (large pulley)	_____	_____	_____

_____ 13. What comparison can be made between the top speed of the permanent split capacitor motor and the

split phase motor? _____

_____

_____ 14. Does the permanent split capacitor motor turn faster? (circle one):   Yes   No

_____ 15. What comparison can be made between the amp draw of the permanent split capacitor motor and

the split phase motor? _____

_____

_____ 16. Does the permanent split capacitor motor draw lower amps? (circle one):   Yes   No

_____ 17. Adjust the permanent split capacitor motor for the blower RPM attained in step 7 from above using

the split phase motor.   Permanent split capacitor amps = _____

_____ 18.   Subtract the permanent split capacitor amps at that speed from the split phase amperage recorded in

step 7.   Difference in amps = _____

_____ 19.   Calculate the seasonal savings for a typical homeowner with a heat only system, assuming 1000
hours of fan operation. Use .8 for the power factor and .08 for the cost per KWH. For example, a
difference of 1 amp will produce a savings of $7.60 in the 1000 hr season.

$$\frac{\text{Difference in amps} \times 0.8 \times 120 \text{ V} \times 1000 \text{ hr} \times \$.080}{1000} = \underline{\hspace{2cm}}$$

INSTALL REPLACEMENT THERMOSTAT

STUDY MATERIAL
Chapter 8, Unit 1

LABORATORY NOTES
This laboratory worksheet will cover the installation of a replacement thermostat. Many thermostats these days are multiday programmable, and may even be multizone programmable. The thermostat you may be changing out could be old enough to have a mercury switch, and should be disposed of at a toxic waste collection site. Your instructor and your employer will have instructions on their procedures for handling mercury switches.

UNIT DATA (HVAC UNIT)

1. Unit description _____

2. Fuel type _____

3. Ignition system _____

4. Control transformer.   VA rating = _____   Location _____

5. Fan relay type and number _____

6. Heating circuit device and rated amp draw _____

7. Cooling capacity _____

8. Number of stages of heat _____

9. Number of stages of cooling _____

10. List other functions to control or supervise _____

THERMOSTAT DATA

1. Make _____   Model # _____

2. Stages of heat _____   Stages of cooling _____

3. Heat anticipator (circle one):   Adjustable   Nonadjustable

4. Cool anticipator (circle one):   Adjustable   Nonadjustable

5. Subbase model # _____

6. Subbase switching (circle all that apply): Heat   Off   Auto   Cool   Fan on   Fan auto

7. List special thermostat features if any. _____

_____

INSTALL THERMOSTAT

Check	Step	Procedure
_____	1.	Check unit operation. If possible, list any problems. _____
_____	2.	Turn off main power to unit.
_____	3.	Check existing thermostat wire condition and size.
		Size (18 gauge is typical) = _____ Number of conductors _____
		Color of wires (circle all that apply): White Red Green Blue Yellow Other
_____	4.	Install and level new thermostat subbase in a stable location.
_____	5.	Match and record thermostat wire color with subbase terminals.
		W1_____ W2_____ O_____ B_____ R or RH_____ RY_____ G_____ Y1_____ Y2_____
_____	6.	Install a jumper from RH to W1.
_____	7.	Turn power to equipment on. Observe heat operation.
_____	8.	Measure amp draw of heating circuit. Amps = _____
_____	9.	Turn power off.
_____	10.	Install thermostat on subbase.
_____	11.	Set heat anticipator to measured amp draw.

CHECK OPERATION

Check	Step	Procedure
_____	1.	Turn thermostat to off position.
_____	2.	Obtain accurate room temperature at thermostat location.
_____	3.	Turn setpoint setting until contacts open.
_____	4.	Compare setpoint with room temperature. Setpoint = _____ Room temperature = _____
_____	5.	Calibrate setpoint to equal room temperature.
_____	6.	Turn power on.
_____	7.	Start and set clock if required.
_____	8.	Use thermostat instructions to program if programmable.
_____	9.	Turn heat-off-cool switch to heat position and turn setpoint to above room temperature.
_____	10.	Observe heat operation.
_____	11.	Turn fan on switch to on position. Observe fan on.

Check	Step	Procedure
_____	12.	Turn fan to auto position. Observe fan off.
_____	13.	Turn heat-off-cool switch to cool position.
_____	14.	Turn setpoint to below room temperature.
_____	15.	Observe cooling operation.
_____	16.	Turn setpoint to above room temperature.
_____	17.	Observe cooling off.
_____	18.	Install thermostat cover.
_____	19.	Demonstrate correct equipment operation to instructor.

TESTING SPLIT PHASE REFRIGERATOR COMPRESSORS

STUDY MATERIAL
Chapter 7, Unit 4

LABORATORY NOTES
This laboratory worksheet will cover the controls and procedures used in testing split phase refrigerator compressors.

OHMS TEST

Check	Step	Procedure
_____	1.	Obtain ohmmeter and a selection of compressors.
_____	2.	Determine compressor to be tested. It must be of the split phase or capacitor start type and 115 V to follow this procedure.
_____	3.	Expose the compressor terminals.
_____	4.	Using the highest ohms scale available on the ohmmeter, test all terminals of compressor to the dome.
_____	5.	Any measurable resistance from the windings to the dome means the compressor is grounded and the test is completed.
_____	6.	If the compressor does not show a ground, proceed with the test and measure terminal to terminal on the R × 1 scale.
_____	7.	Zero the meter, draw a sketch of the terminal location, number the terminals, and record the exact resistance in the space provided.

C = Terminal left over while measuring the highest resistance

S = Common to the higher resistance

R = Common to the lowest resistance

Check	Step	Procedure
		Ohm test 1 1 to 2 = _____ ohms C = _____
		Ohm test 2 2 to 3 = _____ ohms R = _____
		Ohm test 3 1 to 3 = _____ ohms S = _____
_____	8.	Repeat the ohmmeter test as often as required to obtain correct readings.

APPLIED VOLTAGE TEST

Check	Step	Procedure
_____	1.	Obtain a fused start cord approved for the voltage and motor type to be tested.
_____	2.	Perform test on a wooden bench only.
_____	3.	Don't plug in the cord until all other connections have been made and checked.
_____	4.	Measure amperage on the common leg at all times while the motor is energized.
_____	5.	Connect the start cord leads to common, start, and run, according to the color code on the cord.
_____	6.	Connect the two capacitor leads to the start capacitor.
_____	7.	Be sure the rotary switch is in the off position.
_____	8.	Unlock the ammeter needle lock button.
_____	9.	Rotate the rotary scale to the 50 amp scale.
_____	10.	Install the ammeter on the common lead.
_____	11.	Plug in the start cord to 115 V.
_____	12.	Turn the rotary switch to the start position.
_____	13.	Drop back to the run position when compressor operation is verified.
_____	14.	Rotate the ammeter to a lower scale. Record current. Amps = _____
_____	15.	If the compressor fails to run within its rated amps, turn off the compressor at the rotary switch of the start cord.
_____	16.	Repeat the starting procedure as required to verify the compressor operation.
		Second test OK? (circle one): Yes No
		Third test OK? (circle one): Yes No
_____	17.	Obtain a second compressor and repeat the above.
_____	18.	Demonstrate the starting procedure for your instructor.

TESTING PERMANENT SPLIT CAPACITOR AND CAPACITOR START/CAPACITOR RUN COMPRESSOR MOTORS

STUDY MATERIAL
Chapter 7, Unit 4

LABORATORY NOTES
Residential AC units, whether split, package, or self-contained units, use three phase or capacitor motors because of the motor efficiency and the starting power requirements. Both the permanent split capacitor and the capacitor start/capacitor run motors use a run capacitor in series with the start winding at all times making the conventional start cord dangerous to use. The start cord removes the start winding and could damage a permanent split capacitor or a capacitor start/capacitor run motor if it were used to start the motor.

Materials needed: permanent split capacitor motor, no relay or start capacitor required.

OHMS BENCH TEST

Check	Step	Procedure
_____	1.	Obtain ohmmeter, 230 V start cord, an assortment of start and run capacitors, assorted jumper wires, and an ammeter.
_____	2.	Select compressor #1 on test bench. Check with your instructor if you are not sure if compressor is permanent split capacitor type.
_____	3.	Using the highest ohms scale available on the ohmmeter, test all terminals of compressor to the dome.
_____	4.	Determine start, run, and common compressor terminals.

Check	Step	Procedure
_____	5.	Record ohm test results and sketch terminal location in space below.

C = Terminal remaining while measuring highest ohms

S = Terminal of higher resistance from common terminal

R = Terminal of lowest resistance from common terminal

Ohm test 1 to 2 _____   C = _____

Ohm test 2 to 3 _____   R = _____

Ohm test 1 to 3 _____   S = _____

APPLIED VOLTAGE TEST, PERMANENT SPLIT CAPACITOR MOTOR

Check	Step	Procedure
_____	1.	Connect voltage supply wires to C and R. *Note: Do not plug in.*
_____	2.	Hand wire an appropriate run capacitor from S to R terminals. Use manufacturer recommended capacitor or 15 mfd per motor HP rating.
_____	3.	Install ammeter on common.
_____	4.	Energize motor by plugging in or turning on power.
_____	5.	Observe motor operate and amps approach locked rotor amps and drop to full load amps.
_____	6.	Repeat starting process as required to read amperage.
_____	7.	Observe compressor pump air through ports.
_____	8.	Record amps during start in common wire. Amps = _____
_____	9.	Record amps during run in common wire. Amps = _____
_____	10.	Record amps during run in run wire. Amps = _____
_____	11.	Record amps during run in start wire. Amps = _____
_____	12.	Amperage in run wire and common wire should be the same. Did you observe this in your system? (circle one): Yes No
_____	13.	Amperage in wire to start terminal will be about 20% or 30% of common amperage. Did you observe this in your system? (circle one): Yes No

APPLIED VOLTAGE TEST, CAPACITOR START/CAPACITOR RUN

Check	Step	Procedure
_____	1.	Addition of an appropriate start capacitor will convert any permanent start capacitor motor to a capacitor start/capacitor run.
_____	2.	Obtain 230 V start cord and appropriate start capacitor.
_____	3.	Connect 230 V start cord to C, S, and R as indicated on start cord.
_____	4.	Hand wire in run capacitor from S to R terminals. Some 230 V start cords or boxes have both run capacitor and start capacitor connections at the box and do not require hand wiring in the run capacitor. All permanent split capacitor and capacitor start/capacitor run motors require a run capacitor. The start capacitor adds start power, but is not required to run. Do not run any permanent split capacitor or capacitor start/capacitor run motor without a run capacitor in series with the start winding.
_____	5.	Install ammeter on common wire.
_____	6.	Plug in or connect to appropriate voltage supply.
_____	7.	Energize motor with start cord turning rotary switch to start and dropping back to run when the amps drop off and the compressor begins pumping air.
_____	8.	Record amp flow as follows:

	During start	During run
Full load amps at common terminal:	_____	_____
Run winding at R terminal:	_____	_____
Start winding at capacitor:	_____	_____

CHECK/TEST THERMOCOUPLE SYSTEM

STUDY MATERIAL
Chapter 10, Unit 5

LABORATORY NOTES

A thermocouple was that most common pilot supervisory device in the gas heating industry up until the mid 1980s and is still in use today. Some old furnaces from that era are still in use. Some furnaces such as many house trailer heaters, most water heaters, and some space heaters still use them. Typical of technology development, the new stuff doesn't eliminate the old but adds to it. We still have to know how to test, replace, and service problems related to thermocouples.

A good thermocouple is expected to produce 30 mV or .003 V_{DC}. When connected to a gas valve safety switch or separate pilot safety switch, the thermocouple is said to be under load and should produce 9–15 mV. The drop out range of a thermocouple, that is, the point where it is on the verge of failing to hold in the pilot safety valve and turning off the pilot, is considered to be 9–11 mV. This is an operating range and will vary a little. You may find a thermocouple that is producing 10 mV under load but works fine. You may also find some that are nuisance tripping (intermittent pilot outage) at the 12 mV range. The best thing to do in the case of intermittent pilot outage is to perform an under load test and do whatever you can to improve the flame signal output.

THERMOCOUPLE, OPEN CIRCUIT TEST (30 MV)

Check	Step	Procedure
_____	1.	Obtain any standard thermocouple installed or not installed.
_____	2.	Remove the threaded connection from the gas valve or pilot safety switch if necessary.
_____	3.	Connect a mV meter from the outer copper line to the inner core lead at the end of the thermocouple.
_____	4.	Heat the enclosed end of the thermocouple with a torch or a normal pilot assembly.
_____	5.	A voltage of up to 30 mV will be read on the meter. If the meter goes down, reverse the leads.
		Record the highest voltage. Voltage = _____ mV

THERMOCOUPLE, CLOSED CIRCUIT TEST (UNDER LOAD)

Check	Step	Procedure
_____	1.	Obtain a thermocouple installed on a standard combination gas valve or a pilot safety switch.
_____	2.	Obtain the thermocouple adaptor for the valve end.
_____	3.	Light the pilot for a normal pilot flame.
_____	4.	Read the millivolts at the adaptor connection terminals. Voltage = _____ mV
_____	5.	Blow out the pilot light and relight the pilot within 30 sec of going out.
_____	6.	Why does gas still come out of the pilot burner after the pilot is out? _____
_____	7.	Slide the thermocouple up or down until just the top third of the thermocouple is in the flame.
_____	8.	Increase the size of the pilot to obtain the highest voltage reading. Voltage = _____ mV
_____	9.	Remove and scrape the connection mating surfaces in the gas valve safety socket and the rounded end of the thermocouple. A fingernail file works very well for this purpose.
_____	10.	Reposition the pilot assembly as required to obtain the highest voltage reading. Change the thermocouple holder, making a slight bend in a warped flame shield.

Retest the voltage.   Voltage = _____ mV

CHECK/TEST A FLAME ROD SYSTEM

STUDY MATERIAL
Chapter 10, Unit 4

LABORATORY NOTES

A flame rod is a flame supervisory that conducts an electrical signal to the flame and through the flame. A clean gas flame contains ions that will carry electrical current. Because of the difference in surface area of the flame rod and the burner, more electrons will travel in one direction than the other. This process is called flame rectification.

The control box receives an AC power supply that is conducted to the flame rod. When a flame is established, the signal is carried through the flame but rectified in the process. The fact that the flame rod wire carries a DC signal proves that there is a flame present. If the flame rod were to touch the metal of the burner, an AC signal would pass. We can say that a flame is proved or verified by the process of flame rectification.

Positioning of the flame rod for the best signal is critical to the operation of the system. The control module instruction booklet will tell what the signal should be. Flame signals are generally measured in microamps, but sometimes in DC voltage. Sometimes we have to work on a flame rod system without the assistance of the booklet. Generally speaking, if the flame rod signal is steady, and accepted by the module, it is of the correct microamps. Poor quality signals will usually waver and bounce up and down. A consistent signal means a strong signal.

FLAME ROD WITH HOT SURFACE IGNITOR

Check	Step	Procedure
_____	1.	Perform this test on any furnace equipped with a hot surface ignitor and a flame rod for proving flame.
_____	2.	Locate the flame rod, the flame rod wire, and the sensor connector at the burner control module.
_____	3.	Obtain manufacturer booklet and look up the flame signal output. Current = _____ microA
_____	4.	Turn off main power.
_____	5.	Install micrometer in series with flame rod sensor wire to sensor wire terminal of control box.
_____	6.	Turn burner on. Observe normal ignition sequence.
_____	7.	Observe flame. Read flame signal output. Current = _____ microA

Check	Step	Procedure
_____	8.	If flame does not establish or burns poorly refer to normal burner service procedures. Bleed gas line, check adjustment, check manifold pressure, etc.
_____	9.	If flame looks good but signal is poor, remove flame rod and clean with steel wool.
_____	10.	Inspect clean and tighten all connections as required.
_____	11.	Reinstall flame rod and check ignition again.
_____	12.	If flame looks good and flame signal is good, replace the burner control module.

INSTALL, CHECK, AND SET A BOILER RESET CONTROL

STUDY MATERIAL
Chapter 8

LABORATORY NOTES

The boiler reset control is one of the most cost effective controls for saving energy in the typical hot water boiler heating system. Its function is to lower the boiler water operating temperature as the outdoor temperature goes up. This is an inversely proportional relationship. It does this by measuring both outdoor temperature and boiler water temperature. As the outdoor temperature drops, the heat requirement of the building increases, and hotter water is required to maintain building temperature. The outdoor temperature sensing bulb must be placed in a stable outdoor position sensing a good average temperature. Obtain the boiler book from the file and look up the instruction pages for the White-Rogers type 1050 and 1051 indoor-outdoor hot water control. Refer to the booklet for all installation and adjustment questions concerning the control.

VISUAL INSPECTION

Check	Step	Procedure
_____	1.	Perform a visual inspection and identify the following boiler component parts: pump, burner, boiler water inlet, aquastat relay, boiler water outlet, reset control, radiation zone loops.
_____	2.	Locate the burner terminals in the aquastat relay.
_____	3.	Set aquastat relay temperature setting at 200°F for the high limit.
_____	4.	Wire reset control in series with burner.
_____	5.	Perform a mini boiler startup to verify normal operation.
_____	6.	Obtain and read instructions for reset controller.
_____	7.	Set reset controller square dial (reset ratio) setting to 1/1 and round dial to 120°F. The boiler water will now reset 1 degree for each degree of outdoor bulb temperature drop below 70°F with 120°F as the minimum boiler water temperature on any call for heat.
_____	8.	Place outdoor sensing bulb of reset control in room adjacent to boiler, 68/75°F. The boiler water should now seek a minimum water temperature of 120°F.

Check	Step	Procedure

9. Operate boiler with both thermostats calling and both zone valves open until the burner turns on for the second time. This restart temperature should eliminate coast temperature and be the actual minimum opening the valve. Observe water coming out, and the valve close by itself.

Temperature = _____ °F

10. Install the outdoor thermostat in an ice bath and observe the boiler burner come on.

11. Allow burner and pump to operate with both thermostats calling.

12. At second burner on cycle record water temperature.   Temperature = _____ °F

13. Install outdoor thermostat in a freezer at 0°F.

14. Observe burner turn on and off. Record water temperature the second time the burner turns on.

Temperature = _____ °F

15. Test with voltmeter the thermistat of the aquastat relay (L2 of circuit) to R of aquastat relay and then to B.

16. The loss of potential measured on the thermostat of the aquastat relay proves the burner is off on aquastat relay rather than the reset controller.

WIRE FOR ADD-ON COOLING, NEWER PREWIRED FURNACE

STUDY MATERIAL
Chapter 8, Unit 1

LABORATORY NOTES

This laboratory worksheet will be useful for a newer furnace prewired for cooling. Most modern furnaces are equipped for cooling from the factory. This means that the furnace has a larger blower motor, has a fan relay to turn the fan on for cooling mode, and has a furnace mounted terminal strip for control circuit wiring. The industry standard terminal designation is: R = power, C = common, W = heat, Y = cool, G = fan. Some equipment uses V = voltage, H = heat, C = cool, and F = fan. This laboratory worksheet will lead the student through the steps to convert a newer, equipped for cooling furnace that is currently operating on a heat only wiring, to a combination heat/cool system. On a belt drive furnace it is standard to install a larger blower motor and pulley to obtain the larger airflow required for cooling. Oil furnaces would require an isolated heat/cool thermostat subbase to prevent mixing the control circuit of the primary control with the 24 V that runs the cooling system.

OBTAIN AND IDENTIFY ALL PARTS REQUIRED

Check	Step	Procedure
_____	1.	A heat only furnace.
_____	2.	Fan center with 40 VA transformer
_____	3.	Combination Heat/Cool thermostat and subbase.
_____	4.	Sufficient length of four or five wire thermostat wire.
_____	5.	An existing direct drive multispeed blower capable of delivering enough air for the AC tonnage to be installed.
_____	6.	Obtain heat/cool wiring diagram from your instructor, from the furnace literature, or from the furnace.

MAIN POWER WIRING

Check	Step	Procedure
_____	1.	Turn off, lock out, and tag main power to furnace.
_____	2.	Measure voltage to furnace, verify that the voltage is 0 V. V = _____
_____	3.	Mount fan center on a four electrical box. The existing four main power box will be OK if there is one.
_____	4.	Wire neutral (white wire) of fan center to L2 of main power.
_____	5.	Wire black (hot) wire from power supply of furnace to both black of transformer and common side of relay.
_____	6.	Choose manufacturer recommended cooling fan speed by connecting the NO (normally open) relay contacts with wire feeding cooling speed of multispeed blower motors. Check/select also an appropriate heating speed as recommended by the manufacturer.

CONTROL CIRCUIT WIRING

Check	Step	Procedure
_____	1.	Install new heat cool thermostat or heat cool subbase for existing thermostat.
_____	2.	Run new four or five wire thermostat wire, as required, from the furnace subbase to the thermostat.
_____	3.	At thermostat and also at the furnace subbase, connect Red to R, White to W, Green to G, Blue or Yellow to Y.
_____	4.	For oil furnaces only, Red to R, Blue or Yellow to Y, and Green to G.
		_____ a. Run a separate 2 wire for furnace heat mode only.
		_____ b. Connect T and T of primary control to Rc and W.

HEAT/COOL STARTUP

Check	Step	Procedure
_____	1.	Use Laboratory Worksheet GF-1 or O-11 to perform heat system startup.
_____	2.	Use Laboratory Worksheet AC-1 to perform AC system startup.
_____	3.	Have your instructor check over operation of both heating mode and cooling mode.

WIRE FOR ADD-ON COOLING, ELECTRIC SYSTEM

STUDY MATERIAL
Chapter 8, Unit 1

LABORATORY NOTES

This laboratory worksheet will cover the single transformer system that is equipped for cooling. Most modern furnaces are equipped for cooling from the factory. This means that the furnace has a larger blower motor, has a fan relay to turn the fan on for cooling mode, and has a furnace mounted terminal strip for control circuit wiring. The Honeywell standard terminal designation is: R = power, C = common, W = heat, Y = cool, and G = fan. This laboratory worksheet will lead the student through the steps to convert a modern prewired furnace that does not yet have an AC system. It will be assumed that the newer furnace has a direct drive multispeed blower motor. We will wire the blower to operate at a slower speed for heating and a higher speed for cooling.

OBTAIN AND IDENTIFY ALL PARTS REQUIRED

Check	Step	Procedure
_____	1.	A furnace prewired for cooling and equipped with a factory installed subbase (R, W, Y, G, C + etc.).
_____	2.	Combination heat/cool thermostat and subbase.
_____	3.	Sufficient length of four or five wire thermostat wire.
_____	4.	A direct drive multispeed of sufficient airflow capacity to handle the tonnage of AC to be installed.
		Read and record the number of speeds and tonnage of AC recommended.
		Number of speeds = _____ Tonnage = _____
_____	5.	Obtain heat/cool wiring diagram from your instructor, from the furnace literature, or from the furnace.

MAIN POWER WIRING

Check	Step	Procedure
_____	1.	Since the furnace is prewired for cooling there are no main power changes required.
_____	2.	Read and record the number of fan speeds available and which is currently used for heating and for cooling. Total = _____ Heating = _____ Cooling = _____

CONTROL CIRCUIT WIRING

Check	Step	Procedure
_____	1.	Run new thermostat from new subbase to thermostat.
_____	2.	At thermostat connect red to R, white to W, green to G, and blue or yellow to Y.
_____	3.	At furnace connect red to R, white to W, green to G, and blue or yellow to Y.
_____	4.	Run a two wire from the furnace to the condensing unit. Leave a couple feet of extra wire coiled up and out of the way just in case something damages the wire.
_____	5.	Connect the red of the two wire to Y of the furnace and the white of the two wire to C of the furnace.
_____	6.	At the condensing unit, connect the red and white of the two wire to the field connections of the condensing unit or the contactor coil terminals.
_____	7.	Trace/inspect wiring and compare with diagram.
_____	8.	Wiring OK. System ready to start.

CHECK OPERATION

Check	Step	Procedure
_____	1.	Turn on main power to furnace.
_____	2.	Turn thermostat to fan on observe fan come on.
_____	3.	With no main power to condensing unit, turn thermostat to a call for cooling.
_____	4.	Observe contactor coil close for AC operation.
_____	5.	Demonstrate operation for instructor.

WIRE FOR ADD-ON COOLING, GAS FURNACE

STUDY MATERIAL
Chapter 8, Unit 1

LABORATORY NOTES

An older heat only furnace does not have a fan center, fan relay, or cooling condenser fan motor on the blower and these things are installed in the field. The Honeywell standard terminal designation is still the same: R = power, C = common, W = heat, Y = cool, and G = fan. This laboratory worksheet will lead the student through the steps to convert an older heat only gas furnace to a combination heat/cool system. On a belt drive furnace it is standard to install a larger blower motor and pulley to obtain the greater airflow required for cooling.

OBTAIN AND IDENTIFY ALL PARTS REQUIRED

Check	Step	Procedure
_____	1.	A heat only furnace.
_____	2.	Fan center with 40 VA transformer
_____	3.	Combination heat/cool thermostat and subbase.
_____	4.	Sufficient length of four or five wire thermostat wire.
_____	5.	A replacement blower motor or blower assembly capable of delivering the required airflow in units of CFM (if required).
_____	6.	For belt drive blowers only, you will need the following parts: • $1/3$ HP belt drive blower motor • $3^1/2$ to 4 in OD adjustable motor pulley
_____	7.	Obtain heat/cool wiring diagram from your instructor, from the furnace literature, or from the furnace.

MAIN POWER WIRING TO FURNACE

Check	Step	Procedure
_____	1.	Turn main power to furnace off.
_____	2.	Measure voltage to furnace, verify zero voltage. Voltage = _____
_____	3.	Mount fan center on a four wire electrical box. The existing four wire main power box will be OK if there is one.
_____	4.	Wire neutral (white wire) of fan center to L2 of main power.
_____	5.	Wire black (hot) wire from power supply of furnace to both black of transformer and common side of relay.
_____	6.	Install replacement belt drive blower motor (as required).
_____	7.	Connect NO (normally open) relay contacts with wire feeding single speed blower motor or cooling speed on multispeed blower motors.

CONTROL CIRCUIT WIRING

Check	Step	Procedure
_____	1.	Run new thermostat from new subbase to thermostat.
_____	2.	At thermostat connect red to R, white to W, green to G, and blue or yellow to Y.
_____	3.	At furnace fan center at connect red to R, white to W, green to G, and blue or yellow to Y.
_____	4.	Run a two wire thermostat wire from the furnace subbase out to the contactor coil of the AC condensing unit.
_____	5.	Connect the two wire red and white to Y and C of the subbase.
_____	6.	Connect the two wire to the field thermostat connections shown on the condensing unit diagram (contactor coil).
_____	7.	Trace/inspect wiring and compare with diagram.
_____	8.	Wiring OK. System ready to start.

CHECK OPERATION

Check	Step	Procedure
_____	1.	Turn on main power to furnace.
_____	2.	Turn thermostat to fan on. Observe fan come on.
_____	3.	With no main power to condensing unit, turn thermostat to a call for cooling.
_____	4.	Observe contactor coil close for AC operation.
_____	5.	Demonstrate operation for instructor.

WIRE FOR ADD-ON COOLING, OIL FURNACE

STUDY MATERIAL
Chapter 8, Unit 1

LABORATORY NOTES

This laboratory worksheet covers adding wiring for cooling, this time on an oil furnace. Most modern furnaces are equipped for cooling from the factory. This means that the furnace has a larger blower motor, has a fan relay to turn the fan on for cooling mode, and has a furnace mounted terminal strip for control circuit wiring. The Honeywell standard terminal designation is: R = power, C = common, W = heat, Y = cool, and G = fan. This laboratory worksheet will lead the student through the steps to convert an older oil furnace to a combination heat/cool system. On a belt drive furnace it is standard to install a larger blower motor and pulley to obtain the larger airflow required for cooling. Oil furnaces would require an isolated heat/cool thermostat subbase to prevent mixing the control circuit of the primary control with the 24 V that runs the cooling system.

OBTAIN AND IDENTIFY ALL PARTS REQUIRED

Check	Step	Procedure
_____	1.	A heat only furnace.
_____	2.	Fan center with 40 VA transformer.
_____	3.	Combination heat/cool thermostat and subbase prewired with an isolated heat/cool circuit.
_____	4.	Sufficient length of four or five wire thermostat wire.
_____	5.	If required, a replacement blower motor or blower assembly capable of delivering the required airflow in units of CFM.
_____	6.	For belt drive blowers only, you will need the following parts:
		• $1/3$ HP belt drive blower motor
		• $3 1/2$ to 4 in OD adjustable motor pulley
_____	7.	Obtain heat/cool wiring diagram from your instructor, from the furnace literature, or from the furnace.

MAIN POWER WIRING

Check	Step	Procedure
_____	1.	Turn main power to furnace.
_____	2.	Measure voltage to furnace. Verify that the voltage is zero. Voltage = _____
_____	3.	Mount fan center on a four wire electrical box. The existing four wire main power box will be OK if there is one.
_____	4.	Wire neutral (white wire) of fan center to L2 of main power.
_____	5.	Wire black (hot) wire from power supply of furnace to both black of transformer and common side of relay.
_____	6.	Connect NO (normally open) relay contacts with wire feeding single speed blower motor or cooling speed on multispeed blower motors.

CONTROL CIRCUIT WIRING

Check	Step	Procedure
_____	1.	Run new thermostat wire from new subbase to one three or four wire and one two wire for oil primary.
_____	2.	If installed, remove jumper from Rc to Rw.
_____	3.	At thermostat, connect the three or four wire red to Rc, green to G, and blue or yellow to Y.
_____	4.	This step will be found in the oil furnace procedure only. Connect R and W of the two wire to Rc and W.
_____	5.	At the furnace, connect the two wire to T and T of the primary control.
_____	6.	[Optional] Install isolation relay. A standard thermostat can be used on an oil furnace by using an isolation relay. Step 5 above is eliminated and the W wire is brought back to the coil of a 24 V isolation relay. The other side of the relay coil is then wired back to C of the transformer. The NO contacts of the relay are connected to T and T of the primary control, isolating the primary control voltage.

CHECK OPERATION

Check	Step	Procedure
_____	1.	Turn on main power to furnace.
_____	2.	Turn thermostat to fan on. Observe fan come on.
_____	3.	With no main power to condensing unit, turn thermostat to a call for cooling.
_____	4.	Observe contactor coil close for AC operation.
_____	5.	Demonstrate operation for instructor.

WIRE FOR ADD-ON COOLING, POWER PILE/BELT DRIVER

STUDY MATERIAL
Chapter 8, Unit 1

LABORATORY NOTES

A power pile system used to heat a home is becoming a thing of the past. However, every HVAC/R service technician should be familiar enough with such a system to recognize what it is. In any skilled trade you prove yourself not by doing an average job, but by recognizing and solving the problems that others cannot. The power pile system generates its own operating control circuit voltage from the series of thermocouple connections called the power pile. Its advantage is that you can have building heat even when the power goes off. You cannot, however, have any blower operation without main power, nor could you have AC. Occasionally people do have a power pile and still have a forced air system and add AC to the forced air system.

However, mixing the control circuit voltage of the power pile with the 24 V that runs the cooling system cannot be done. The 24 V of the normal control circuit will burn up the power pile circuitry. The power pile only generates 250 to 750 millivolts (.25–.75 V) and requires a special gas valve and thermostat to run it. The anticipator of a standard thermostat would cause too much voltage drop.

OBTAIN AND IDENTIFY ALL PARTS REQUIRED

Check	Step	Procedure
_____	1.	A heat only furnace with a power pile, pilot assembly, gas valve, and thermostat.
_____	2.	Fan center with 40 VA transformer.
_____	3.	Combination heat/cool subbase matching the power pile thermostat with an isolated heat/cool circuit.
_____	4.	Sufficient length of three or four wire thermostat wire and also some two wire thermostat wire.
_____	5.	If required, a replacement blower motor or blower assembly capable of delivering the required airflow in units of CFM.
_____	6.	For belt drive blowers only, you will need the following parts:
		• $1/3$ HP belt drive blower motor
		• $3\frac{1}{2}$ to 4 in OD adjustable motor pulley
_____	7.	Obtain heat/cool wiring diagram from your instructor, from the furnace literature, or from the furnace.

MAIN POWER WIRING

Check	Step	Procedure
_____	1.	Run main power to furnace using a handy box and SSU switch. Leave turned off until wiring is completed.
_____	2.	Measure voltage to furnace. Verify that the voltage is zero. Voltage = _____
_____	3.	Mount fan center on a four wire electrical box. The existing four wire main power box will be OK if there is one.
_____	4.	Wire neutral (white wire) of fan center to L2 of main power.
_____	5.	Wire black (hot) wire from power supply of furnace to both black of transformer and common side of relay.
_____	6.	Connect NO (normally open) relay contacts with wire feeding single speed blower motor or cooling speed on multispeed blower motors.

CONTROL CIRCUIT WIRING

Check	Step	Procedure
_____	1.	Run new thermostat wire from new subbase to one three or four wire.
_____	2.	Remove jumper from Rc to Rw (if installed).
_____	3.	At thermostat connect the 3 or 4 wire red to Rc, green to G, and blue or yellow to Y.
_____	4.	Connect R and W of the two wire to Rc and W.
_____	5.	At the furnace connect the two wire to the connections of the old thermostat wire at the gas valve circuit. Trace control wires and be sure the furnace limit control is in series with the gas valve.
_____	6.	[Optional] Connect an isolation relay. A standard thermostat can be used on a millivolt system by using an isolation relay. Step 5 from above is eliminated and the W wire is brought back to the coil of a 24 V isolation relay. The other side of the relay coil is then wired back to C of the transformer. The NO contacts of the relay are connected to the gas valve and furnace limit control circuit.

CHECK OPERATION

Check	Step	Procedure
_____	1.	Turn on main power to furnace.
_____	2.	Turn thermostat to fan on. Observe the fan come on.
_____	3.	With no main power to condensing unit, turn thermostat to a call for cooling.
_____	4.	Observe contactor coil close for AC operation.
_____	5.	Demonstrate operation for instructor.
_____	6.	Light pilot and measure voltage output of power pile. Voltage = _____ mV
_____	7.	Turn thermostat to a call for heating.
_____	8.	Observe a normal heat mode:
		_____ a. Burner on
		_____ b. Fan on

DRAW SYSTEM SCHEMATIC

STUDY MATERIAL
Chapter 8, Unit 3

LABORATORY NOTES
The purpose of this lab is to get the student used to sketching wiring diagrams in the field from the actual unit wiring. In reality the service technician will do a considerable amount of troubleshooting without the aid of a factory wiring diagram. In addition, there may have been unit modifications rendering the factory diagram incomplete. It is to the advantage of the technician to construct a field wiring diagram as a service aid to avoid confusion when troubleshooting complex panels thus saving time and costly mistakes.

UNIT DATA

1. Select any shop built system; AC, Refrigeration, Heat/Cool, Boiler. List or describe system.

    _____

2. Describe function of equipment. _____

    _____

3. Make a list of every electrical component, categorized by loads and by switches. Select and draw the symbol to be used for that component in the diagram to be drawn.

	Loads	Symbol	Switches	Symbol
1.	_____	____	_____	____
2.	_____	____	_____	____
3.	_____	____	_____	____
4.	_____	____	_____	____
5.	_____	____	_____	____

4. First draw a pictorial diagram of the system. Draw the symbol of the component in the approximate location and connect the wires as they are connected in the system. Show color coding.

5. Convert the diagram to a schematic format. Put L1 on the left, and L2 on the right. Connect the load unbroken to L2. Select location of switches in order from left to right.

PROGRAM A THERMOSTAT

STUDY MATERIAL
Chapter 8, Unit 1

LABORATORY NOTES

Energy conservation is playing an increasingly more important part in today's heating and air conditioning systems. The electronic programmable clock thermostat is becoming a popular energy management tool for the residential and light commercial market. This device allows the customer to automatically change set point temperatures for occupied times and set back or set up the unoccupied temperature depending on heating mode or cooling mode.

Since each of these operate and program a little differently, keeping the instruction booklet is essential. Certain features have common functions but the function name may vary from brand to brand. Some are full seven day programmable while some are 5 day/2 day clocks, with all weekdays treated the same and both weekend days treated the same. A business would usually require a full 7 day programmable thermostat. Typically there are a maximum of four daily times (wake, leave, come home, and sleep). The four times are designed for two setback periods; one during the night and one during the day when people are at work. This type of thermostat can be used for business applications by setting the midday setback at one temperature all day or stacking the middle time points on one point in time. Various manufacturers of thermostats will use different terms to describe these and other features of the thermostat. This laboratory worksheet will provide some practice in programming each type.

5 DAY/2 DAY THERMOSTAT OR 5 DAY/1 DAY/1 DAY

Check	Step	Procedure
_____	1.	Obtain a 5 day/2 day or a 5 day/1 day/1 day, with four daily setback periods. Your instructor will provide the thermostat for you.
_____	2.	Read and follow the manufacturer's instructions.
_____	3.	Set clock to correct time of day.
_____	4.	Set the 5 day (weekday) setting for:

6 AM (wake), set to 70°F

7:30 AM (leave), set to 62°F

4:30 PM (home), set to 70°F

10 PM (sleep), set to 60°F

Check	Step	Procedure
_____	5.	Set the 2 day (weekend) setting for:

       8 AM (wake), set to 70°F

       11 AM (leave), set to 65°F

       5 PM (home), set to 70°F

       11:30 PM (sleep), set to 60°F

FULL 7 DAY PROGRAMMABLE

Check	Step	Procedure
_____	1.	Obtain a 7 day programmable thermostat.
_____	2.	Read and follow the manufacturer's instructions.
_____	3.	Set clock to correct time of day.
_____	4.	Follow the manufacturer's instructions to program the thermostat for a typical retail store.

       9 AM to 8 PM Monday through Thursday

       9 AM to 11 PM on Friday

       8 AM to 11:30 PM Saturday

       9 AM to 11 PM Sunday

The system will be in the heating mode. Provide a 60°F temperature for unoccupied time, 70°F for occupied time, and allow lead time for morning warmup based on a 5°F/hr pick-up, meaning that at 8 AM the temperature should be set to 65°F. The heat could be turned off 1/2 hr before closing time and the building would still coast or be comfortable.

GAS FURNACE LABORATORIES

GAS FURNACE STARTUP

STUDY MATERIAL
Chapter 10, Unit 1

LABORATORY NOTES
This laboratory worksheet will cover the normal service of gas furnace startup.

UNIT DATA

1. Furnace make _____ Model # _____

2. Furnace type (circle one): Upflow Counterflow Basement Horizontal

3. Blower type (circle one): Direct drive Belt drive

4. Blower motor speeds (circle one): Single speed 2 3 4

5. System type (circle one): Heat only Heat and humidify Heat and cool

6. Burner type (circle one): Atmospheric Induced draft Power

PRESTART CHECKS

Check	Step	Procedure
_____	1.	Vent connector connected with three screws per joint.
_____	2.	Fuel line installed properly, with no apparent leaks.
_____	3.	Spin all fans to be sure they are loose and spin freely.
_____	4.	Combustion area free from debris.
_____	5.	Electrical connections complete. Main Power _____ Thermostat _____
_____	6.	All doors and panels available and in place.
_____	7.	Thermostat installed and operating correctly.

START UP AND CHECK OPERATION

Check	Step	Procedure
_____	1.	Turn power off and thermostat below room temperature.
_____	2.	Measure voltage at plug duplex to be used. Voltage = _____
_____	3.	Check fuse with ohmmeter record Amperage rating. Amps = _____
_____	4.	Plug unit in and turn power on.
_____	5.	Obtain fan only operation. To do this, use thermostat fan on, or push button of fan limit, or consult instructor.
_____	6.	Light pilot as required.
_____	7.	Turn fan off and thermostat to heat.
_____	8.	Adjust temperature setting to above room temperature by 10°F.
_____	9.	Observe flame sequence begin and flame on.
_____	10.	Observe blower begin after heat exchanger warms up.
_____	11.	Turn thermostat down and observe flame off.
_____	12.	Observe blower off within 3 min of burner off.

READINGS AND MEASUREMENTS

Check	Step	Procedure
_____	1.	Obtain normal heating operation.
_____	2.	Measure and record. Discharge air temperature = _____ Room air temperature = _____
_____	3.	Calculate temperature rise by the following formula:

Discharge air temperature – Room air temperature = Temperature rise = _____

Check	Step	Procedure
_____	4.	Read rated burner output from nameplate. Btu = _____
_____	5.	Use the following formula to calculate airflow by temperature rise method:

$$\text{CFM} = \frac{\text{Burner output from step 3}}{1.08 \times \text{Temperature rise from step 4}} = \underline{\quad} \text{ CFM}$$

Check	Step	Procedure
_____	6.	Measure fan motor amperage. Amps = _____
_____	7.	Read rated fan motor amperage from nameplate or motor. Amps = _____
_____	8.	Blower motor amperage should be lower than the blower motor rating.
_____	9.	Call your instructor over to demonstrate furnace operation and explain any problems or missing parts.

MEASURE GAS USAGE

STUDY MATERIAL
Chapter 10, Unit 1

UNIT DATA

1. Unit make _____   Unit model # _____

2. Burner make _____   Burner type _____

3. Number of main burner orifices = _____

4. Number of pilot orifices = _____

5. Appliance input rating in Btu/hr (also known as BTUH) = _____

CHECK FURNACE OPERATION
This follows a generic procedure, not any one particular brand of furnace.

Check	Step	Procedure
_____	1.	Light pilot as required.
_____	2.	Turn gas valve to on.
_____	3.	Turn power on.
_____	4.	Turn thermostat up.
_____	5.	Observe burner on.
_____	6.	Observe fan on.
_____	7.	Turn down thermostat.
_____	8.	Observe burner off.
_____	9.	Observe fan shut down.

CLOCK MAIN BURNER

Check	Step	Procedure
_____	1.	Isolate main burner of appliance to be tested.
_____	2.	Leave pilot on for normal operation.
_____	3.	Operate main burner to be tested.
_____	4.	Observe $1/2$ ft^3 or 2 ft^3 dial of gas meter rotating.
_____	5.	Measure time in seconds for one revolution of dial.
_____	6.	Record number of cubic feet of gas from dial and number of seconds from watch. (circle one):

$1/2$ ft^3 2 ft^3 Time = _____ sec

Check	Step	Procedure
_____	7.	Use the following formula and measured data from #6 to obtain actual gas burned:

$$CFH = \frac{3600}{\text{Number of seconds}} \times \text{Cubic feet} = \underline{\hspace{1cm}}$$

Check	Step	Procedure
_____	8.	Actual burner input is equal to gas consumed times 1000 Btu, or contact utility to verify heating value of gas.
_____	9.	Burner input in Btu/hr. Heat input = _____ Btu/hr
_____	10.	Compare actual measured input with the rated firing rate on nameplate.
_____	11.	If there is a difference of greater than 5%, measure and record the gas manifold pressure. OK?

(circle one):   Yes   No   Pressure = _____ in WC

Check	Step	Procedure
_____	12.	Divide the firing rate by the number of gas orifices to obtain orifice Btu size. Btu = _____

CLOCK PILOT BURNER

This procedure will cover the 5 min method.

Check	Step	Procedure
_____	1.	Obtain a pilot only operation.
_____	2.	Record gas consumed in 5 min by counting the dial movement and the number of marks on the meter face. (Mark to mark = .05 ft^3)
_____	3.	Pilot gas in 5 min is equal to: _____ = _____ ft^3
_____	4.	Use the following formula to calculate the pilot gas consumed in a 30 day month, a 200 day season and a 365 day year:

(5 min pilot gas) $\times$ 12 $\times$ 24 hr $\times$ 30 day = _____

(5 min pilot gas) $\times$ 12 $\times$ 24 hr $\times$ 200 day = _____

(5 min pilot gas) $\times$ 12 $\times$ 24 hr $\times$ 365 day = _____

Check	Step	Procedure

_____ 5. Calculate cost of pilot gas consumed by using the cubic feet of gas consumed from #4 above times the current cost per 100 ft³ of gas.

One month:	Cost per 30 day pilot = _____
One heating season:	Cost per 200 day pilot = _____
One year:	Cost per 365 day pilot = _____

CLOCK PILOT BURNER (short cut method, 100 ft. units)

Check	Step	Procedure

_____ 1. To determine the cost of a 30 day pilot, time the number of seconds mark to mark on the ½ ft³ dial during a normal pilot only operation. Insert the seconds into the following formula:

$$\text{Cost} = 1296 \times \frac{\text{Unit cost}}{\text{Number of seconds}} = \underline{\hspace{1cm}}$$

The unit cost is the gas cost per 100 ft³. 1296 is a conversion factor.

GAS FURNACE PREVENTIVE MAINTENANCE (PM)

STUDY MATERIAL
Chapter 10, Units 2 & 3

LABORATORY NOTES
This laboratory worksheet covers gas furnace preventive maintenance.

UNIT DATA

1. Make _____   Model # _____

2. Furnace type (airflow design) _____

3. Blower type (circle one):   Direct   Belt

4. Number of blower speeds = _____

5. Btu input rating = _____   Number of burner sections = _____

6. Calculate the number of Btu per burner on clamshell type heat exchangers.   Btu = _____

PRESTART CHECKS

Check	Step	Procedure
_____	1.	Consult with customer for any known problems.
_____	2.	Perform visual inspection for the following:

    _____ a. Vent connector complete and meets code.

    _____ b. Vestibule free from combustion material.

    _____ c. Vestibule free from soot or burned wires.

    _____ d. No loose or dangling wires or components.

    _____ e. All covers and panels in correct position.

    _____ f. Furnace in apparent operable condition.

    _____ g Notify instructor or homeowner about any apparent malfunctions.

INITIAL STARTUP

Check	Step	Procedure
_____	1.	Turn fused power supply or SSU switch on if required.
_____	2.	Obtain fan only operation, verify power to furnace.
_____	3.	Turn thermostat to highest setting.
_____	4.	Observe a normal sequence of operation (burner on, fan on, burner off, fan off).
_____	5.	Inspect operating pilot and comment on pilot color, size, shape, position, etc.
_____	6.	Observe several burner ignitions. Use fused power supply switch or SSU switch.
_____	7.	Comment on burner light off concerning lifting, floating, noise, and smooth or even ignition.
_____	8.	Observe continuous main burner operation closely.
_____	9.	Verify burning rate by clocking gas meter. Airflow = _____ CFM
_____	10.	Measure and record gas manifold pressure in units of inches water column.
		Pressure = _____ in WC
_____	11.	Observe blower startup.
_____	12.	Comment on belt tension, bearing noise, dirt in blower blades.

BLOWER MAINTENANCE

Check	Step	Procedure
_____	1.	Turn power off. Test with meter to verify.
_____	2.	Remove wires from blower motor at an accessible location.
_____	3.	Remove screws or bolts securing blower assembly.
_____	4.	Inspect belt for cracks and signs of wear.
_____	5.	Inspect pulleys for wear, grooving, and alignment.
_____	6.	Spin blower by hand, observe pulleys turn.
_____	7.	Inspect for pulley wobble and alignment.
_____	8.	Listen for bearing noise, drag, or movement.
_____	9.	Inspect blower shaft for signs of wear.
_____	10.	Clean blower and motor with air pressure, brushes, scrapers, and cleaning solution as required.
_____	11.	Lubricate motor and blower bearings with oil as required.
_____	12.	Reassemble blower, taking care to check pulley alignment and correct belt tension. Check with instructor for correct blower operation.
_____	13.	Do not install blower until heat exchanger has been checked.

BURNER MAINTENANCE

Check	Step	Procedure
_____	1.	Remove and clean pilot assembly.
_____	2.	Clean and inspect flame sensor and igniter.
_____	3.	Remove and clean main burners, inspect and mark burners for original location. All burners are not interchangeable.
_____	4.	Remove vent connector, draft diverter, and any flue baffles.
_____	5.	Vacuum and brush all soot, rust, and solid particles from the fire side of the heat exchanger.
_____	6.	Insert a light into combustion area.
_____	7.	If possible turn off lights in furnace room.
_____	8.	With the light in each burner, inspect the heat exchanger from both the fan side and the plenum side for holes.
_____	9.	Reinstall blower assembly after the heat exchanger check.
_____	10.	Reinstall main and pilot burners.
_____	11.	Reconnect vent components.
_____	12.	Install a gas pressure gauge on main burner manifold.
_____	13.	Light pilot and adjust for proper size, configuration, location, and positioning.
_____	14.	Perform pilot turndown test as required.
_____	15.	Turn on main burner. Observe operation.
_____	16.	Adjust gas pressure to 3.5 in WC.
_____	17.	Adjust air shutter to obtain correct flame color and CO_2.
_____	18.	Perform post combustion efficiency test at this time. Do this step only if pre and post combustion testing are required.
_____	19.	Adjust fan control for correct fan operation.
_____	20.	Unplug or disconnect blower.
_____	21.	Observe burner operate with no fan until burner turns off, (burner cycles on limit). Perform limit check.
_____	22.	Reinstall all panels.
_____	23.	Check final operation on thermostat.
_____	24.	Demonstrate correct operation for customer. Be ready to "Show and Tell" any additional problems noted or corrected with equipment.

GAS COMBUSTION TESTING

STUDY MATERIAL
Chapter 10, Unit 3

LABORATORY NOTES

Combustion testing is primarily used when setting power burners on which there is an adjustment that will allow total overcombustion to the burner. The traditional standard setting is at 50% excess air to ensure complete combustion. Consult the manufacturer recommendation for the burner, furnace or boiler type for more accurate settings. When using one of the newer electronic combustion analyzers, the appearance of CO will show when complete combustion of the fuel is taking place.

A modern induced draft burner is designed to get plenty of air for combustion and cannot be changed enough to make any difference. They typically operate at 75% excess air but maintain thermal efficiency by getting all the heat back out of the excess air with the larger heat exchanger. Condensing furnaces will frequently test out with a higher thermal efficiency than their combustion efficiency. The flue gasses are so cold that they can condense in the combustion testing equipment and cause problems with the equipment.

Perfect combustion of a fuel is accomplished by having exactly the right amount of air mixed exactly with the fuel to burn the fuel. This mixture is dangerous to strive for since too little oxygen will waste fuel and produce carbon monoxide. There are no simple methods to test for unburned fuel in the vent gases. The ultimate CO_2 giving a perfect combustion mixture for the various fuels is listed below. Due to the difficulty of ensuring a complete mixing, it is best to burn gas with about 50% excess air, this will ensure all the fuel is burned, with a minimum of stack loss. The percentage will vary more for natural gas because the consistency of natural gas varies as to amounts of liquid petroleum, air, and water in the gas. Natural gas means *natural as found* and is not pure methane. Oil and propane are more pure compounds and burn rates are more predictable. There will not be much difference in CO_2 when performing this test on atmospheric and induced draft burners because of the limited control over the combustion air intake.

Ultimate CO_2	Recommended CO_2		
(Perfect Combustion)	25%	50%	75%
	Excess Air	Excess Air	Excess Air
(Approximate Values)			
Natural Gas = 11.7 – 12.2	9%	8%	7%
Propane = 13.7	11.5%	9.5%	8%
Number 2 Oil = 14.7	12.5%	10.5%	9%

153

Ultimate CO_2	Recommended CO_2		
(Perfect Combustion)	25%	50%	75%
	Excess Air	Excess Air	Excess Air
(Approximate Values)			
All Fuels = 0%	5%	7%	9%

UNIT DATA

1. Shop ID # _____
2. Furnace make _____   Model # _____
3. Burner type (circle one):   Atmospheric   Induced draft   Power
4. Input to burner = _____ CFH

PREPARATION FOR COMBUSTION TESTING

Check	Step	Procedure
_____	1.	Inspect furnace for operable condition.
_____	2.	Vent complete.
_____	3.	Fuel line complete and leak free.
_____	4.	No loose wires dangling unconnected.
_____	5.	All panels in place or adjacent to furnace.
_____	6.	Thermostat installed and operable.
_____	7.	Locate CO_2 test openings and temperature probe location for undiluted vent gas sample.
_____	8.	Insert thermometer probe and check CO_2 probe for position.
_____	9.	Turn on manual main power.
_____	10.	Light pilot as required.
_____	11.	Turn on main power and turn thermostat for a call for heat.
_____	12.	Observe burner on and fan on.
_____	13.	Observe continuous burner operation until vent gas temperature tops out.

COMBUSTION TEST

Check	Step	Procedure
_____	1.	Take initial readings and record in space provided below.
_____	2.	Close primary air shutter until a lazy yellow flame is present. Test the yellow flame.
_____	3.	Open the air shutter to a maximum, test with excess air.
_____	4.	Close the air shutter to obtain 8% CO_2.

Check	Step	Procedure

_____ 5.   Leave the burner set at 8% CO_2.

_____ 6.   Using the slide rule calculator with the natural gas slide, look up the combustion efficiency.

Combustion Test Readings	Test 1 (initial)	Test 2 (yellow)	Test 3 (excess)	Test 4 (8% CO_2)
CO_2	_____	_____	_____	_____
Actual stack temperature	_____	_____	_____	_____
Net stack (gross – room)	_____	_____	_____	_____
Combustion efficiency	_____	_____	_____	_____

7.   Conclusions. Which flame is the most efficient? _____

_____

Other comments. _____

_____

MEASURE FURNACE THERMAL EFFICIENCY

STUDY MATERIAL
Chapter 10, Unit 3

LABORATORY NOTES
The purpose of this test is to measure thermal efficiency of a furnace and to draw a correlation, if there is any, between the airflow through a furnace and the thermal efficiency of the furnace. Thermal efficiency is Btu output divided by Btu input. Input is a measure of gas consumed or fuel consumed, while output is measured by the sensible heat formula, temperature × 1.08 × airflow. The same furnace will be used to perform three operating tests and therefore the thermal efficiency is calculated at three different airflows.

UNIT DATA

1. Furnace make _____   Model # _____

2. Shop ID # _____   Fuel _____

3. Burner type (circle one):   Power   Atmospheric   Induced draft

4. Rated input in Btu/hr _____   Return air size _____

5. Estimated thermal efficiency by generic furnace/burner type.

   a.   Condensing = 92%

   b.   Induced draft = 84%

   c.   Typical power burner = 80%

   d.   Standing pilot with electronic ignition and vent device = 75%

   e.   Standing pilot with just electronic ignition = 70%

   f.   Standing pilot operating as manufactured = 60%

PRETEST OPERATION

Check	Step	Procedure
_____	1.	Obtain normal heating operation.
_____	2.	Observe system operation with a continuous burner operation.
_____	3.	Slow down blower speed or close off registers to obtain the lowest possible airflow through furnace and still maintain 100% burner operation. *Note: Burner must not cycle on limit during test.*

TEST DATA
Record data below.

Check	Step	Procedure
_____	1.	Obtain fan only operation. Airflow hood will be damaged if exposed to temperatures above 140°F.
_____	2.	Measure furnace airflow on inlet and/or outlet using airflow hood. It may be necessary to build a bracket for furnace to install hood. Airflow = _____ CFM
_____	3.	Remove airflow hood from discharge of furnace.
_____	4.	Operate furnace in normal heating mode. Airflow hood may be left on return air during this test. Airflow will typically go down as the air is heated up due to the air restriction of the furnace and the increased air volume of the heated air.
_____	5.	Measure temperature rise through furnace. Temperature rise = _____
_____	6.	Calculate furnace Btu output by the following formula:

Air temperature rise × Airflow × 1.08 = Btu

_____ × _____ × 1.08 = _____

| _____ | 7. | Determine Btu input by one of the three methods listed here. |

a. Clock meter.

b. Burner rated input.

c. Actual nozzle size (oil only).

| _____ | 8. | Determine thermal efficiency of the furnace by the following formulas: |

Btu output / BTU input = % thermal efficiency

Test 1 (Initial airflow): _____/_____ = _____%

| _____ | 9. | Lower the furnace airflow and repeat the above test. |

Test 2 (Reduced airflow): _____/ _____ = _____%

Check	Step	Procedure
_____	10.	Raise the furnace airflow and repeat the above test.

Test 3 (Increased airflow): _____/_____ = _____%

Check	Step	Procedure
_____	11.	What correlation can be drawn between the airflow and the thermal efficiency of the furnace?

_____

Check	Step	Procedure
_____	12.	As the airflow increases, the thermal efficience will

(circle one):   Increase   Decrease   Not change

BASIC GAS FURNACE REPLACEMENT

STUDY MATERIAL
Chapter 10

LABORATORY NOTES
The simplest gas furnace replacement assumes that the old gas furnace has failed and is not worth repairing. It is a replacement of the furnace only without any duct system changes or AC considerations. This is usually done through one of the following three options, first a *give me your best price*, second a *get it done today emergency situation*, or third a *warranty replacement*. The old furnace is disconnected, pulled out and the new one is slid under the old duct connection and hooked back up.

REMOVAL OF EXISTING GAS FURNACE

Check	Step	Procedure
_____	1.	Shut off electrical power at fuse box, remove fuse.
_____	2.	Shut off gas at meter.
_____	3.	Disconnect gas line and twist the support and plug out of the way as required. Note that pipe that is removed and stacked on the floor becomes used pipe and cannot be used without inspection.
_____	4.	Turn off power to furnace, tag and lock out at main box.
_____	5.	Disconnect and remove SSU (power supply switch) and conduit from furnace.
_____	6.	Disconnect vent pipe from furnace to chimney.
_____	7.	Clean up basement before installation of new furnace.

INSTALLATION OF NEW GAS FURNACE

Check	Step	Procedure
_____	1.	Remove furnace from carton.
_____	2.	Measure and cut in location of return air filter fitting. Install filter fitting.
_____	3.	Slide furnace under old return air and supply air plenum.
_____	4.	Install return air plenum fitting. Rivet or screw to furnace.

Check	Step	Procedure
_____	5.	Install air filter.
_____	6.	Move furnace around for the simplest connection of supply air plenum, return air plenum, gas line, and electrical connections.
_____	7.	Measure and make connection to supply air plenum.
_____	8.	Measure and make connection to return air plenum.
_____	9.	Connect existing gas line to furnace with code required shutoff, union, and drip leg.
_____	10.	Install new vent and vent connector as per manufacturer recommendation.
_____	11.	Install barometric damper in a location recommended by the furnace manufacturer.
_____	12.	Connect main power supply to furnace and verify ground polarity. Local code may require a licensed electrician for electrical connection.
_____	13.	Connect thermostat wire to furnace terminal strip as per wiring diagram.
_____	14.	Check/replace thermostat as required.
_____	15.	Set thermostat anticipator at manufacturer recommended setting.
_____	16.	Connect furnace condensate outlet to a suitable drain or to a condensate pump.
_____	17.	Turn gas on and check for any gas leaks.
_____	18.	Turn thermostat to a call for heat.
_____	19.	Observe draft inducer on, normal flame on, and indoor fan on—in that sequence.
_____	20.	Observe a normal heat mode on and cycle off.
_____	21.	Measure and record fan motor amperage. Amps = _____
_____	22.	Measure and record operating temperature difference of heat mode.

Temperature difference = _____ |
| _____ | 23. | Demonstrate normal operation to the customer. |

INSTALL A GAS FURNACE AND DUCT SYSTEM

STUDY MATERIAL
Chapters 10 & 14

LABORATORY NOTES

This lab details the steps generally followed to do a complete gas furnace installation including a total revamp of the air duct system. Older systems typically had branch ducts originating from a heat only plenum based duct system, inadequate by today's standards. Systems today require a larger size duct system that is ready for central AC. Modern furnaces are shorter than the old furnace being replaced and a matching cased coil is sometimes installed to gain height, even though AC is not put in at that time, just to match the height of the old system.

Some of the old duct may be reused, but it is frequently full of dirt from years of use. Slow air velocities and poor filters over years of use will allow dirt to accumulate, especially in the return air ducts. Frequently the entire duct system is replaced to solve these problems. Some companies own their own duct cleaners or work with a duct cleaning company. It is a given to evaluate and either clean or replace the entire duct system in a complete upgrade.

REMOVAL OF EXISTING GAS FURNACE

Check	Step	Procedure
_____	1.	Inspect for asbestos duct wrapping. If asbestos is present obtain a licensed asbestos abatement company to remove the old furnace and all ducting.
_____	2.	Shut off electrical power at main electrical box. Install lockout tag and lock at main power supply box.
_____	3.	Shut off gas at meter.
_____	4.	Bleed gas from gas pipe.
_____	5.	Dismantle gas pipe, disassemble to straight lengths of pipe and single fittings.
_____	6.	Remove electrical conduit from furnace to junction box or main breaker terminal.
_____	7.	Disconnect vent pipe from furnace to chimney.
_____	8.	Cover register openings in house.
_____	9.	Remove runouts from plenum and at square to round fittings. Save usable pipe for reuse.
_____	10.	Dismantle old furnace and remove from basement.
_____	11.	Remove all old panning.

_____ 12.   Remove or have cleaned any existing air ducts that are to be connected to or used in new system.

_____ 13.   Clean up basement before installation of new furnace.

INSTALLATION OF HIGH EFFICIENCY GAS FURNACE

Check	Step	Procedure
_____	1.	Select location of furnace.
_____	2.	Remove furnace from carton.
_____	3.	Measure and cut in location of return air filter fitting. Install filter fitting.
_____	4.	Install return air plenum fitting and pop rivet or screw to furnace. Install air filter.
_____	5.	Install matched cased coil, if supplied.
_____	6.	Install supply air plenum.
_____	7.	Install transitional offset(s) to supply air plenum.
_____	8.	Snap chalk lines on joists where supply and return truck lines will run.
_____	9.	Locate return air registers, nail panning to joists, and install headers.
_____	10.	Measure and cut in any top takeoffs for branch ducts.
_____	11.	Install supply air truck line to transitional offset fitting, and secure to joists.
_____	12.	Cut in and install side takeoffs for supply registers. Install enough warm air pipe to clear other side of return air trunk line.
_____	13.	Cut in holes in return air trunk line, and install to return air plenum.
_____	14.	Locate all return air panning, measure 1 in inside the edge where return air trunk line will be located and cut in holes for return air.
_____	15.	Measure from panning to return air plenum, and cut in holes in return air trunk line.
_____	16.	Install return air trunk line and secure to joists.
_____	17.	Finish installing all supply warm air pipe to square to round fittings.
_____	18.	Use an approved duct sealing caulk to close all significant duct air leaks.
_____	19.	Complete gas pipe connection as per code.
_____	20.	Turn on gas at meter, observe an initial rush of gas through the meter to fill the system and then 0 gas flow.
_____	21.	Bleed gas from line at appliance union.
_____	22.	Inspect all gas lines and connections for leaks.
_____	23.	Install power supply SSU switch with 15 amp fuse to furnace.
_____	24.	Install conduit and run electrical wires to junction box, or main breaker terminal.
_____	25.	Test and shut off electrical power.
_____	26.	Run new six conductor thermostat wire to thermostat location, and install thermostat.
_____	27.	Remove covers from register openings in house.
_____	28.	Install vent pipe as per manufacturer recommendation. Use chimney liners when venting 80+ furnaces into conventional brick/block and mortar chimneys.
_____	29.	Most modern 90+ furnaces require a two pipe vent system using 2 in or larger PVC.

Check	Step	Procedure

_____ 30.   Install ventilation air and combustion air intakes as recommended by the manufacturer or local code.

_____ 31.   Remove lockout tag and lock.

_____ 32.   Turn on electrical power.

_____ 33.   Check operation of furnace.

_____ 34.   Measure and record the following:

      a. Gas pressure in units of in WC to operating appliance.   Pressure = _____ in WC

      b. Temperature rise of air through the furnace.   Temperature rise = _____

      c. Amps of fan motor.   Amps = _____

_____ 35.   Clean up basement.

_____ 36.   Demonstrate and instruct customer in operation of new furnace and thermostat.

Total hours to complete: _____

Time and materials list would require a complete list of hours and materials used.

INSTALL COMPLETE GAS FURNACE, AC, AND DUCT SYSTEM

STUDY MATERIAL
Chapter 10

LABORATORY NOTES

This lab details the ultimate goal of every heating/cooling shop when a furnace replacement job is in order. This gives the customer the highest level of comfort and control while giving the contractor the full assurance that the entire system has been upgraded. Additional options to the new furnace, AC system, and upgraded duct system are generally a humidifier, electronic air cleaner, programmable thermostat, multiple zoning, other control options, and any air exchange ventilation system. There seems to no end to the new options and devices on the market for the new generation of educated demanding customers.

Frequently some changes will be made in register locations to improve air circulation to certain areas of the building or to supply air to a new addition, etc. This type of job often is chosen as a part of a complete remodeling project.

REMOVAL OF EXISTING GAS FURNACE

Check	Step	Procedure
_____	1.	Shut off electrical power at fuse box, and remove fuse.
_____	2.	Shut off gas at meter.
_____	3.	Bleed gas from gas pipe.
_____	4.	Dismantle gas pipe, disassemble to straight lengths of pipe and single fittings.
_____	5.	Remove electrical conduit from furnace to junction box, or main breaker terminal.
_____	6.	Disconnect vent pipe from furnace to chimney.
_____	7.	Cover register openings in house.
_____	8.	Remove runouts from plenum and at square to round fittings. Save usable pipe for reuse.
_____	9.	Dismantle old furnace and remove from basement.
_____	10.	Remove old panning. This step is a necessity for gravity systems. Forced air furnaces may not require removal of panning.
_____	11.	Inspect and remove if dust buildup is visible. Remove or have cleaned any existing air duct.
_____	12.	Clean up basement before installation of new system.

INSTALLATION OF EQUIPMENT

Check	Step	Procedure
_____	1.	Remove selected old registers as required.
_____	2.	Cut in new register openings as required.
_____	3.	Select location of furnace.
_____	4.	Remove furnace from carton.
_____	5.	Measure and cut in location of return air filter fitting. Install filter fitting. Even if not to be installed at the current time, try to leave room for an electronic air filter.
_____	6.	Install return air plenum fitting and pop rivet or screw to furnace. Install air filter.
_____	7.	Install matched cased coil.
_____	8.	Install supply air plenum.
_____	9.	Locate and install AC condensing unit.
_____	10.	Cut holes in wall and run line set. Be careful not to kink suction line or rip insulation.
_____	11.	Leak test line set and evacuate as required. Use other installation time to leak test and evacuate.
_____	12.	Using field duct design procedures, layout and properly size a duct distribution system that will distribute sufficient air for the AC capacity being installed.
_____	13.	Allow for current or future multizoning in duct design as required.
_____	14.	Install transitional offset(s) to supply air plenum.
_____	15.	Snap chalk lines on joists where supply and return truck lines will run.
_____	16.	Locate return air registers, nail panning to joists, and install headers.
_____	17.	Measure and cut in any top takeoffs for branch ducts.
_____	18.	Install supply air trunk line to transitional offset fitting, and secure to joists.
_____	19.	Cut in and install side takeoffs for supply registers. Install enough warm air pipe to clear other side of return air trunk line.
_____	20.	Complete all supply air main and branch duct.
_____	21.	Cut in holes in return air trunk line, and install to return air plenum.
_____	22.	Locate all return air panning, measure 1 in inside the edge where return air trunk line will be located, and cut in holes for return air.
_____	23.	Install return air trunk line and secure to joists.
_____	24.	Install return air mixing box and return air drop including connection with filter slot, electronic air cleaner, or special filter as required.
_____	25.	Seal all plenum and duct corners with duct sealing caulk.
_____	26.	Complete gas piping. Turn on gas at meter, check for leaks, and bleed gas from line.
_____	27.	Install SSU switch with 15 amp fuse to furnace.
_____	28.	Install conduit and run electrical wires to junction box, or main breaker terminal.
_____	29.	Test and shut off electrical power. Code may require a licensed electrician for electrical connection.
_____	30.	Run new six conductor thermostat wire to thermostat location, and install thermostat.

Check	Step	Procedure
_____	31.	Install vent pipe as per manufacturer recommendation. Use chimney liners when venting 80+ furnaces into conventional brick/block and mortar chimneys.
_____	32.	Most modern 90+ furnaces require a two pipe vent system using 2 in or larger PVC.
_____	33.	Install ventilation air and combustion air intakes as recommended by the manufacturer or local code.
_____	34.	Install humidifier as supplied.
_____	35.	Turn on electrical power.
_____	36.	Check operation of furnace and AC system.
_____	37.	Perform Laboratory Worksheet AC-2, Complete AC Startup, on system.
_____	38.	Clean up basement.
_____	39.	Demonstrate and instruct customer in operation of new system and thermostat.

Comments: _____

_____

Total hours to complete: _____

A time and materials sheet would require a complete list of hours and materials used.

INSTALL A GAS CONVERSION BURNER

STUDY MATERIAL
Chapter 10, Unit 3

LABORATORY NOTES

Gas conversion burners are atmospheric inshot type burners that are designed to fit in the place of a high pressure gun type oil burner. Power gas burners are also made that will fit. The furnace or boiler must have a drum type heat exchanger with a single combustion chamber. This type of heat exchanger is required for oil burners and is common in mobile home furnaces. They can be set up for either natural gas or LPG. Natural gas will use 3.5 in WC for the gas pressure and LPG uses 11 in WC.

Since we have total control over the gas air ratio, it is recommended we use an electronic combustion analyzer to set the air to the fuel we are burning by either the CO_2 or the O_2 readout. The correct procedure is to set the fuel burning rate to the appliance Btu input and set the air to match that. The natural gas burning rate can be double checked by clocking the meter and frequently these types of burners have adjustable orifices. LPG systems usually have a replaceable fixed orifice and you must choose the one required. Proper setup of the burner is extremely important as any field changes to the furnace bring the liability to the person working or the company the person works for.

UNIT DATA

1.   Shop ID # _____

2.   Unit make _____   Model # _____

3.   Furnace type (airflow design) _____

4.   Blower type (circle one):   Direct   Belt

5.   Number of blower speeds = _____

6.   Btu input rating = _____   Number of burner sections = _____

PRESTART CHECKS

Check	Step	Procedure
_____	1.	Consult with customer for any known problems.
_____	2.	Remove thermostat wires from oil burner primary control.
_____	3.	Obtain air circulation fan only operation.

Check	Step	Procedure
_____	4.	Measure and record fan motor amperage. Amps = _____
_____	5.	Read and record nameplate amperage of fan motor. Full load amps = _____
_____	6.	Perform an initial heat exchanger inspection.
_____	7.	Notify customer of any problems with blower or heat exchanger before proceeding with job.

REMOVE EXISTING OIL BURNER

Check	Step	Procedure
_____	1.	Turn off, tag, and lock out main power.
_____	2.	Obtain bucket or pan and rags to catch spilled oil.
_____	3.	Turn thermostat to highest setting.
_____	4.	Disconnect and remove oil lines.
_____	5.	Cap or plug oil fittings on oil pump.
_____	6.	Remove 120 V power and thermostat line from burner.
_____	7.	Remove bolts from burner mount flange and remove burner.
_____	8.	Remove and discard all old vent piping.

CLEAN UP AND INSPECT HEAT EXCHANGER

Check	Step	Procedure
_____	1.	Remove and clean blower assembly.
_____	2.	Insert trouble light into combustion chamber.
_____	3.	Remove panels as required to expose heat exchanger.
_____	4.	Inspect heat exchanger thoroughly for any signs of rust or deterioration.
_____	5.	Thoroughly clean the existing chimney.

INSTALL NEW GAS BURNER

Check	Step	Procedure
_____	1.	Read burner instructions and determine how to obtain correct firing rate in Btu/hr.
_____	2.	Hold burner in position to mark mounting screw holes. Use oil burner mount bolts as available.
_____	3.	Mount burner in position.
_____	4.	Route and run gas line as per code.
_____	5.	Wire according to manufacturer supplied wiring diagram.
_____	6.	Install new thermostat as required.

Check	Step	Procedure
_____	7.	Set heat anticipator at recommended amperage.
_____	8.	Install blower and reconnect wiring.
_____	9.	Install new vent connector according to burner manufacturer instructions and code requirements.
_____	10.	Install all panels for normal furnace operation.

STARTUP AND CHECK OPERATION OF NEW BURNER

Check	Step	Procedure
_____	1.	Bleed gas line at burner drip leg or union.
_____	2.	Remove lockout tag and lock. Turn on power at main box.
_____	3.	Light pilot as required.
_____	4.	Turn power on at SSU.
_____	5.	Turn thermostat to a call for heat.
_____	6.	Observe trial for ignition begin. Keep combustion chamber visible from a reasonable distance and off slightly to one side for safety in case of flame rollout.
_____	7.	Observe burner come on within 1 min.
_____	8.	Observe a normal main flame.
_____	9.	Adjust air intake initially by hand until a clean smooth flame is apparent.
_____	10.	Measure time in seconds for one revolution of the 2 ft³ gas dial. One revolution = _____ sec
_____	11.	Divide 3600 by number of seconds from above and multiply by 2 to get ft3/hr input, sometimes referred to as CFH. Input = _____ ft³/hr (or CFH)
_____	12.	Multiply CFH input from 11 by 1,000 (or local heating value of natural gas) to get heating value in Btu/hr (sometimes referred to as BTUH). Heating value = _____ Btu/hr (or BTUH)
_____	13.	Check with manufacturer instructions and make required adjustments if burner differs more than 5% from furnace manufacturer recommended firing rate.
_____	14.	Observe indoor blower come on when heat exchanger warms up.
_____	15.	Turn thermostat off and observe normal cycling.

FINAL ADJUSTMENTS AND SETTINGS

Check	Step	Procedure
_____	1.	Obtain a combustion analyzer to check combustion.
_____	2.	Drill required holes in vent pipe for probe.
_____	3.	Obtain a normal heat mode.
_____	4.	Allow 5 min warmup time.

Check	Step	Procedure
_____	5.	Record initial CO_2 reading.　　$CO_2 = $ _____
_____	6.	Adjust combustion air shutter to obtain an 8% CO_2 reading for 50% excess air.
_____	7.	Allow 5 min continued operation.
_____	8.	Read and record final CO_2 reading.　　$CO_2 = $ _____

BELT DRIVE BLOWER, COMPLETE SERVICE

STUDY MATERIAL
Chapter 10, Unit 3

LABORATORY NOTES

Some older residential and most commercial air handlers use belt drive blowers to deliver air through the duct system. The advantage of belt drive blowers over direct drive is the flexibility in choosing a blower speed and the ease of manufacturing a blower wheel to withstand the RPM. It is too difficult to match all the factors, RPM, CFM, motor HP, etc., in all commercial applications. Most new residential systems have gone over to direct drive blowers but in commercial systems belt drives will be around for many years to come. Occasionally a belt drive blower needs a complete overhaul including any or all of the following: motor and bearings, motor pulley, belt, blower pulley, blower bearings, blower shaft, cleaning, balancing, or rebalance.

UNIT DATA

1. Unit name _____ Unit model # _____ Serial # _____

2. Unit type _____

3. Blower motor data: Voltage = _____ HP = _____ Locked rotor amps = _____

   Rated load amps = _____ Full load amps = _____

4. Blower wheel data: Width = _____ Diameter = _____ Shaft Size OD = _____

5. Pulley size: Motor pulley OD_____/ID_____ Blower shaft OD_____/ID_____

6. Belt size = _____ Belt number = _____

7. Lubrication required for blower bearings (circle one): Grease Oil

8. Lubrication required for motor bearings (circle one): Grease Oil

9. Inspect greased motors for grease relief fitting at bottom of bearing below grease opening. None _____

ORIGINAL OPERATION/INSPECTION

Check	Step	Procedure
_____	1.	Spin blower wheel slowly by hand.
_____	2.	Note any drag noise or pulley wobble.

Check	Step	Procedure
_____	3.	Install ammeter on L1 to motor.
_____	4.	Place blower door in normal position for normal airflow and load on motor.
_____	5.	Obtain fan only operation, record motor amps. Amps = _____
_____	6.	Remove blower door for further inspection.
_____	7.	Observe motor begin to turn and watch for belt slipping during startup.
_____	8.	Note any running noise, bearing, grinding or any noise other than normal airflow noise.
_____	9.	Spin blower slowly for inspection. Observe pulleys and shaft turning slowly for any wobble or distortion.
_____	10.	Watch belt and pulleys during startup for any slipping, jumping, or abnormal movement.

REMOVAL AND CLEANING

Check	Step	Procedure
_____	1.	Turn off, tag, and lock out power at main disconnect.
_____	2.	Measure voltage L1 and L2, L1 to ground, etc. Voltage is off.
_____	3.	Remove and store carefully all panels and covers.
_____	4.	Loosen tension on belt for removal.
_____	5.	Remove belt and inspect for any cracks or severe shine (caused by slipping).
_____	6.	Inspect pulleys for any wear grooves.
_____	7.	Inspect blower wheel blades and scrape with a screwdriver or thin tool for any evidence of accumulated debris.
_____	8.	Clean blower wheel with water, compressed air, or CO_2, whichever is most available and appropriate.
_____	9.	Blow motor air passages with compressed air or CO_2.
_____	10.	Wipe motor and nameplate clean with a rag. Motor needs to be clean for cooling.

DISASSEMBLY AND INSPECTION

Check	Step	Procedure
_____	1.	Loosen setscrew from blower pulley and remove pulley.
_____	2.	Support blower wheels to keep them in position.
_____	3.	Loosen screws on set collars and pull blower shaft out of blower.
_____	4.	Record length and diameter of blower shaft. Length = _____ OD = _____
_____	5.	List any special features of blower keyways, flat sides, etc, for replacement. _____
_____	6.	Comment on blower shaft condition as far as wear, grooving, rust, etc. _____
_____	7.	Roll shaft on a flat surface to be sure it is straight and true.

Check	Step	Procedure
_____	8.	Remove and inspect or replace blower wheel bearings.
_____	9.	Install original or replacement bearings in blower housing.
_____	10.	Install original or replacement shaft with set collars and spacers to ensure smooth operation.
_____	11.	Install original or replacement shaft pulley.
_____	12.	Install original or replacement motor and pulley.
_____	13.	Align or check alignment of pulleys with a straight edge on outer edge. Straight edge should touch on outer edge of both pulleys, four places total.
_____	14.	Lubricate motor and shaft bearings as required.
_____	15.	Install new belt of correct size. Belt size = _____
_____	16.	Adjust motor tension assembly for correct tension. Belt play of 2–3 in is average. Consult manufacturer of blower for application and recommended tension.
_____	17.	Replace any loose or worn components of belt tension assembly as required.

CHECK FOR CORRECT OPERATION OF BLOWER AND MOTOR

Check	Step	Procedure
_____	1.	Write with a permanent marker, new motor amperage in a service accessible location.
_____	2.	Turn on main power.
_____	3.	Install ammeter on L1 or wire going to C of motor.
_____	4.	Obtain fan only operation.
_____	5.	Put all panels in normal position to obtain normal airflow.
_____	6.	Measure and record motor full load amps with all panels in position. Measured amps = _____
_____	7.	Compare measured amps with rated motor amps.
		Actual amps is (circle one): Higher Lower
_____	8.	Loosen the motor belt tension and remove belt.
_____	9.	Loosen and adjust motor pulley or replace with a pulley of larger or smaller size as required. Move pulley sheave closer to increase size.
_____	10.	Install belt and adjust belt tension.
_____	11.	Run blower and measure amps. Amps = _____
_____	12.	Repeat steps 8–11 to obtain rated full load amps and maximum CFM.

PILOT TURNDOWN TEST

STUDY MATERIAL
Chapter 10, Unit 4

LABORATORY NOTES
A pilot turndown test is a test to determine the smallest possible pilot capable of proving flame presence to the control circuit and lighting the main burner safely. It is performed by lighting the pilot, measuring the pilot flame signal, and observing a series of main burner ignitions using various pilot sizes. Remember a cold burner is more difficult to ignite and in the performance of this test the burner becomes warmed up; you must allow sufficient cooldown time to perform the final test. A pilot turndown test will also point out other pilot problems.

THERMOCOUPLE, OPEN CIRCUIT TEST (30 MV)

Check	Step	Procedure
_____	1.	Obtain any standard thermocouple installed or not installed.
_____	2.	Remove the threaded connection from the gas valve or pilot safety switch if necessary.
_____	3.	Connect a millivolt meter from the outer copper line to the inner core lead at the end of the thermocouple.
_____	4.	Heat the enclosed end of the thermocouple with a torch or a normal pilot assembly.
_____	5.	A voltage of up to 30 millivolts will be read on the meter. If the meter goes down reverse the leads.
		Record the highest millivoltage. Voltage = _____ mV

THERMOCOUPLE, CLOSED CIRCUIT TEST

Check	Step	Procedure
_____	1.	Obtain a thermocouple installed on a standard combination gas valve or a pilot safety switch.
_____	2.	Obtain the thermocouple adaptor for the valve end.
_____	3.	Light the pilot for a normal pilot flame.
_____	4.	Read the millivolts at the adaptor. Voltage = _____ mV
_____	5.	Blow out the pilot light and relight the pilot within 30 sec of going out.

Check	Step	Procedure
_____	6.	Why does gas still come out of the pilot burner after the pilot is out? _____
		_____
_____	7.	Blow out pilot again and observe voltage output.
_____	8.	Record the voltage in millivolts when the pilot valve closes (pilot will not relight) stopping pilot gas (or the pilot safety switch snaps open). The value should be between 9 and 12 mV.
		Voltage = _____ mV
_____	9.	Adjust the size of the pilot flame for the smallest pilot flame capable of proving the pilot and lighting the main burner smoothly and safely. Voltage = _____ mV

FLAME ROD PILOT SYSTEM (REFERENCE = FURNACE BURNER BOOKLET)

Check	Step	Procedure
_____	1.	Perform this test on any furnace equipped with a pilot, a separate spark igniter, and a flame rod.
_____	2.	Locate the pilot adjustment screw. This is usually a needle valve screw located under a screw cap labeled pilot adj.
_____	3.	Turn off main power.
_____	4.	Install microammeter in series with flame rod sensor wire to sensor wire terminal of control box.
_____	5.	Turn on power, and obtain a pilot only flame. Turn off manual main burner valve or pull wire labeled main at redundant gas valve. Consult wiring diagram as required to locate main gas wire.
_____	6.	Temporarily wire a small toggle switch in series with main burner wire and main gas valve. Turn switch off.
_____	7.	Obtain operation of pilot only flame.
_____	8.	Turn on main burner toggle and observe main burner on.
_____	9.	Read flame signal of pilot only in microamps. Current = _____ microA
_____	10.	Turn system off.
_____	11.	Remove and clean flame rod with steel wool.
_____	12.	Read flame signal of pilot only in microamps. Current = _____ microA
_____	13.	Has signal changed from step # 6 above? (circle one): Yes No
_____	14.	Reposition in flame to obtain a better contact with clean blue flame and a stronger signal, if possible.
_____	15.	Turn pilot adjustment screw and observe pilot flame and corresponding microA signal get smaller.
_____	16.	Obtain smallest flame possible capable of proving pilot flame. Current = _____ microA
_____	17.	Turn on main burner toggle, observe main burner on.
_____	18.	Observe several main burner ignitions.
_____	19.	Enlarge pilot as required to provide smooth main burner ignition. Current = _____ microA

FLAME/IGNITOR SYSTEM (REFERENCE = FURNACE BURNER BOOKLET)

Check	Step	Procedure
_____	1.	Perform this test on any furnace equipped with a pilot and a combination pilot ignitor/pilot proving device.
_____	2.	Locate the pilot adjustment screw. This is usually a needle valve screw located under a screw cap labeled pilot adj.
_____	3.	Turn off main power.
_____	4.	Install appropriate toggle switch in series with main gas terminal of redundant gas valve.
_____	5.	Turn on power, and obtain a pilot only flame.
_____	6.	Turn pilot adjustment valve in (clockwise, cw), observe pilot get smaller.
_____	7.	Continue turning in until the pilot goes out and pilot ignitor goes out.
_____	8.	Turn pilot adjustment out (ccw) and observe pilot relight.
_____	9.	With pilot on, turn on main gas toggle and observe main gas burner turn on.
_____	10.	With main burner on, adjust pilot smaller.
_____	11.	Observe main burner go off when pilot goes out.
_____	12.	Adjust pilot to size that will light easily and ignite main burner consistently.

SEPARATE PILOT GAS PRESSURE REGULATOR

This section deals with any commercial burner controller utilizing a separate pilot gas pressure regulator, such as a Honeywell commercial RA89F series controller. This procedure will use a plug in flame monitor jack or wire meter in series with the flame rod.

Check	Step	Procedure
_____	1.	Use any commercial power burner.
_____	2.	Turn off power, main gas valve, and pilot gas valve.
_____	3.	Obtain a microammeter and a plug in flame monitor jack or wire microammeter is series with flame rod.
_____	4.	Remove primary control cover.
_____	5.	Open pilot gas valve and turn on power.
_____	6.	Observe pilot flame only.
_____	7.	Record pilot microamp reading. Current = _____ microA
_____	8.	Decrease gas pressure by turning gas pressure regulator ccw (out). Observe pilot flame get smaller and flame signal reduce in current (microamps).
_____	9.	Turn pilot down until controller relay opens. Current = _____ microA
_____	10.	Increase pilot size until controller relay closes. Current = _____ microA
_____	11.	Turn on main burner gas and power.
_____	12.	Observe main burner ignition.
_____	13.	Adjust pilot as required to obtain smooth main burner ignition during cold start.
_____	14.	Put away all tools and test equipment.

CHECK/TEST/REPLACE HOT SURFACE IGNITER

STUDY MATERIAL
Chapter 10, Unit 4

LABORATORY NOTES

Hot surface ignition and flame proving is the most common method of burner control in modern gas furnaces. The big replacement item in this system is the hot surface igniter. This lab will lead you through a complete check of the ignition system operation and verification of a defective igniter.

Hot surface igniters are sometimes called glow coils because they glow when energized. That is the first tipoff that an igniter has failed, it does not glow when energized. A typical startup sequence will be: thermostat calls for heat, draft fan comes on, hot surface igniter glows, gas valve opens (you can hear the click), gas ignites, gas flame reaches flame rod and proves, and normal heat mode is in progress. When the glow coil fails to glow, you hear the click of the gas valve opening but get no flame; chances are you have a failed hot surface igniter.

UNIT DATA

1. Shop ID # _____

2. Make _____ Model # _____

IDENTIFY THE FOLLOWING COMPONENTS OR SYSTEMS

Check	Step	Procedure
_____	1.	Draft inducer fan and induced draft vent system.
_____	2.	Hot surface ignitor.
_____	3.	Flame rod for flame proving system.
_____	4.	Electronic module for flame system control.
_____	5.	Locate and write down the terminal on the module that the flame rod wire connects to. _____
_____	6.	Locate the plug or wire connection from the control system to the hot surface igniter (glow coil).
_____	7.	Is the connection a plug, wire nut connection, or other? (circle one): Yes No
		Describe. _____
		_____

OBSERVE A NORMAL TRIAL FOR IGNITION SEQUENCE

Check	Step	Procedure
_____	1.	Turn thermostat to a call for heat.
_____	2.	Observe draft inducer fan come on.
_____	3.	Does hot surface igniter glow? Check about 1 min after fan on. (circle one): Yes No
_____	4.	If hot surface ignitor does not get hot, did you hear the gas valve click as it opened? (circle one): Yes No
_____	5.	Assuming Yes to steps 3 and 4, turn power off, isolate the hot surface igniter and perform a continuity test on the coil.
_____	6.	Is their continuity through the hot surface igniter? (circle one): Yes No
_____	7.	If No to step #6, condemn it and replace it. If Yes, continue.
_____	8.	Install voltmeter in plug feeding hot surface igniter.
_____	9.	Observe another trial for ignition.
_____	10.	Read and record voltage to the igniter during the middle of the trial for ignition sequence. Volts = _____
_____	11.	If a voltage is read, turn off, plug HSI into plug, and repeat trial for ignition test.
_____	12.	Obtain a pull trigger portable propane soldering torch or extended gas grill type lighter.
_____	13.	Obtain another trial for ignition sequence.
_____	14.	Hold lighted flame at position of hot surface igniter.
_____	15.	When gas valve clicks, flame should ignite normally.
_____	16.	With flame in progress, remove flame rod wire from ignition module terminal.
_____	17.	Observe flame go out immediately.
_____	18.	You have now verified that the glow coil is defective, but that once the flame is on it will burn normally and be protected by the normal flame sensing device. Any loss of flame signal will turn the system off.
_____	19.	Reignite the flame as desired to warm the house for one cycle only and obtain a new glow coil. Prepare to install.

OBTAIN AND INSTALL NEW GLOW COIL

Check	Step	Procedure
_____	1.	Obtain replacement glow of same configuration on both the shape of the coil and connection. A flat igniter can sometimes be replaced with a round one.
_____	2.	When an igniter has the same configuration but a different plug end, complete the following steps, if necessary.
		_____ a. Remove original plug from original igniter with special plug tool made for this. See your instructor for this tool.

Check	Step	Procedure

_____   b. As a last resort, cut the original plug and wire nut. Install a new glow coil using high temperature wire nuts.

_____   3.   Turn furnace off and install new glow coil.

_____   4.   Be sure to install any shield supplied to protect coil.

CHECK NORMAL IGNITION SEQUENCE AND HEAT CYCLE

Check	Step	Procedure

_____   1.   Turn power on and obtain normal heat mode.

_____   2.   Observe burner on, flame on, fan on sequence.

_____   3.   Pull flame rod wire off and observe flame off.

_____   4.   Reconnect flame rod wire and observe furnace.

_____   5.   Does furnace go into a trial for ignition by itself after the simulated flame failure?

(circle one):   Yes   No

_____   6.   If No, turn power off and back on. Observe startup.

_____   7.   Some furnaces will go into a continuous indoor fan operation after a flame failure. Does yours do this? (circle one):   Yes   No

_____   8.   Some furnaces will have to be on at the furnace 120 V, and turned off and then back on at the thermostat, to end a flame a flame failure lockout. Does yours do this?

(circle one):   Yes   No

_____   9.   Observe and record the process for ending a flame failure mode. _____

_____

_____   10.   Be prepared to demonstrate this to your customer (your instructor).

OIL FURNACE LABORATORIES

OIL FURNACE STARTUP

STUDY MATERIAL
Chapter 11, Unit 1

LABORATORY NOTES
This laboratory worksheet will cover the basics of starting up an oil furnace at the beginning of the heating season.

UNIT DATA

1. Shop ID# _____

2. Furnace make _____   Model # _____

3. Furnace type (circle one):   Upflow   Basement   Counterflow   Horizontal

PRESTART CHECKS

Check	Step	Procedure
_____	1.	Vent connector complete with three screws per joint.
_____	2.	Fuel line complete, installed properly, and leak free.
_____	3.	Spin fan blower, inspecting belt, pulley, and bearings for signs of wear, pulley grooving, or belt cracks.
_____	4.	Combustion area free from debris.
_____	5.	All electrical connections complete, thermostat, and main power.
_____	6.	All doors and panels in place.
_____	7.	Thermostat installed correctly and operating.
_____	8.	Fuel oil present in tank.
_____	9.	Barometric damper installed and swinging freely.

START UP AND CHECK OPERATION

Check	Step	Procedure
_____	1.	Turn power off and thermostat below room temperature.
_____	2.	Plug in cord and then turn on power supply switch or SSU switch.
_____	3.	Obtain fan only operation. Use thermostat auto/on or fan limit manual button.
_____	4.	Observe normal operation of blower.
_____	5.	Turn blower off and thermostat to a call for heat position.
_____	6.	Observe burner on.
_____	7.	Observe fan on after a reasonable warmup time.
_____	8.	Turn down thermostat, observe burner off, and fan off, in that order.
_____	9.	Turn on and off for several ignitions.
_____	10.	The flame should be orange/white in color, uniform in shape and within the combustion chamber, quiet, quick in ignition and extinction, and with no "post nasal drip."
_____	11.	Comments on flame. _____

TAKE READINGS AND MEASUREMENTS

Check	Step	Procedure
_____	1.	Obtain normal heating operation.
_____	2.	Measure discharge air temperature. Temperature = _____°F
_____	3.	Determine airflow by temperature rise method shown below:

$$\text{Airflow} = \text{Burner output}/1.08 \times \text{Temperature difference} = \underline{\hspace{1cm}} \text{ CFM}$$

Check	Step	Procedure
_____	4.	Measure fan motor amps. Amps = _____
		Is this within motor rating? (circle one): Yes No
_____	5.	Check amp rating of oil primary control. Amps = _____
_____	6.	Remove thermostat cover and check/set heat anticipator setting to primary control amp.
_____	7.	Remove and put away all test equipment and tools. Install all panels and doors. Pull plug and hang cord on holder.

OIL BURNER TUNEUP

STUDY MATERIAL
Chapter 11, Units 2 & 3

LABORATORY NOTES
This job is to be completed after the initial oil burner startup job. This laboratory worksheet, as well as the startup laboratory worksheet, refers primarily to a high pressure gun, the most common burner type. I suggest that you complete this job on a different oil furnace than LP1 for a better variety of experience on oil burners.

UNIT DATA

1. Shop ID# _____

2. Furnace make _____   Model # _____

3. Furnace type (circle one):   Upflow   Basement   Counterflow   Horizontal

4. Burner make _____   Model # _____

5. Recommended nozzle data.   GPH _____   Spray pattern _____   Spray angle _____

PRESTART CHECKS

Check	Step	Procedure
_____	1.	Check with customer for any known problems.
_____	2.	Inspect the following:

    _____ a. Vent connector complete.

    _____ b. Check for fuel leaks.

    _____ c. Remove dirt and debris.

    _____ d. Oil filter installed.

    _____ e. Filters installed.

    _____ f. Inspect blower/belt.

    _____ g. Barometric moves freely.

    _____ h. Fans spin freely.

START UP AND CHECK OPERATION

Check	Step	Procedure
_____	1.	Stat calls for heat.
_____	2.	Observe burner on.
_____	3.	Observe fan on.
_____	4.	Turn stat off.
_____	5.	Observe burner off.
_____	6.	Observe fan off.
_____	7.	Demonstrate for customer any major problems with burner.

BURNER INSPECTION AND TUNEUP

Check	Step	Procedure
_____	1.	Turn off furnace at power supply switch or SSU switch at furnace.
_____	2.	Open or remove transformer.
_____	3.	Remove nozzle assembly.
_____	4.	Record nozzle data. Make _____ GPH _____ Angle _____ Pattern _____
_____	5.	Install oil pressure gauge in oil supply line.
_____	6.	Obtain watch with second hand.
_____	7.	Remove and clean cad cell.
_____	8.	Measure and record resistance of cad cell while the face of the cell is covered. Resistance = _____ Ohms
_____	9.	Measure and record resistance of cad cell while the face of the cell is exposed to room light. Resistance = _____ Ohms
_____	10.	With cad cell removed, turn burner on and observe trial for ignition.
_____	11.	Time out and record seconds for burner run time. Time = _____ sec
_____	12.	Wait 2 min for safety switch heater to cool.
_____	13.	Plug in cad cell and leave exposed to light.
_____	14.	Reset safety switch heater.
_____	15.	Observe dark start feature. The burner will not run.
_____	16.	Turn power supply switch or SSU switch off and remove yellow cad cell wires from primary control. The yellow wires can be found at primary control terminals F and F.
_____	15.	Obtain a 1200 ohm resistor.
_____	16.	Install one end of the 1200 ohm resistor to one F terminal at the primary control.
_____	17.	Turn on power supply switch or SSU switch and observe burner run with no oil leaks or spark.
_____	18.	Connect 1200 ohm resistor to second F terminal of primary.

192

Check	Step	Procedure
_____	19.	Observe burner run beyond trial for ignition time measured in step #11 from above.
_____	20.	Adjust oil pressure to 100 PSIG or manufacturer recommended pressure. Record final oil pressure. Pressure = _____
_____	21.	Remove 1200 ohm resistor.
_____	22.	Observe burner turn off due to flame failure within time.
_____	23.	Remove and clean electrodes as required.
_____	24.	Replace nozzle with recommended nozzle from Unit Data.
_____	25.	Install and adjust electrodes.
_____	26.	Position electrodes to manufacturer recommended setting or use the dimensions 1/2 in above, 1/16 in forward, and 1/8 in apart.
_____	27.	Install nozzle assembly.
_____	28.	Slide nozzle to midpoint of forward/backward adjustment.
_____	29.	Swing up or reinstall ignition transformer.
_____	30.	Make sure transformer contact springs touch electrodes as the transformer is positioned.
_____	31.	Secure transformer in place with at least one screw for testing.
_____	32.	Turn on burner and observe ignition. Do not get too close as oil burners can puff back.
_____	33.	Observe a normal continuous burn for 1 min.
_____	34.	Adjust air shutter back and forth slowly, close air to see smoke, and open until smoky flame just disappears.
_____	35.	Slide nozzle assembly forward and back until flame cleans up and quiets down for best flame.

FINAL BURNER ADJUSTMENTS

Check	Step	Procedure
_____	1.	Obtain a draft gauge, smoke gun, and CO analyzer to perform final settings on oil burner.
_____	2.	Use the draft gauge and adjust the barometric damper to obtain a −.01 over fire draft.
_____	3.	Measure and record draft at breach of burner. A restriction of greater than −.04 through the heat exchanger means the heat exchanger may be plugged with soot or rust and flakes and needs internal cleaning. Final measured draft at breach = _____
_____	4.	Inspect and clean fire side of heat exchanger as required.
_____	5.	Use the smoke gun to measure a # 0 smoke reading.
_____	6.	Install CO meter into breach of furnace.
_____	7.	Readjust air shutter to obtain 10% CO_2.
_____	8.	Turn burner on and off, observing several ignitions.
_____	9.	Verify a quick clean quiet ignition.
_____	10.	Set thermostat anticipator at primary control low voltage amp rating.
_____	11.	Remove and put away all test equipment and tools. Install all panels and doors. Pull plug and hang cord on holder.

OIL FURNACE PREVENTIVE MAINTENANCE

STUDY MATERIAL
Chapter 11, Units 2 & 3

LABORATORY NOTES
This laboratory worksheet covers typical oil furnace preventive maintenance.

UNIT DATA

1. Shop ID # _____

2. Unit make _____ Model # _____

3. Furnace type (airflow design) _____

4. Blower type (circle one): Direct belt

5. Number of belt speeds = _____

6. Burner/furnace recommended nozzle data: GPH _____ Spray pattern _____ Spray angle _____

7. Burner/furnace recommended oil pressure. Pressure = _____ PSIG

8. Actual nozzle flow rate in GPH. _____

   Btu/hr input _____ × .8 = Btu output = _____ (Use 140,000 Btu/hr per gallon.)

PRESTART CHECKS

Check	Step	Procedure
_____	1.	Consult with customer for any known problems.
_____	2.	Perform visual inspection for the following parts
	_____ a.	Vent connecter complete and safe.
	_____ b.	Vestibule free from combustion material.
	_____ c.	Vestibule free from soot or burned wires.
	_____ d.	No loose or dangling wires or components.

Check	Step	Procedure
_____	e.	All covers and panels in correct position.
_____	f.	Furnace in apparently operable condition.
_____	g.	Notify instructor or homeowner about any apparent malfunctions noticed.

INITIAL STARTUP

Check	Step	Procedure
_____	1.	Turn on power supply switch or SSU switch if required.
_____	2.	Obtain fan only operation, verify power to furnace.
_____	3.	Turn thermostat to highest setting.
_____	4.	Observe the normal sequence of operation as follows. Burner on, fan on, burner off, fan off.
_____	5.	Inspect operating flame and comment on flame color, size, shape, position, etc.
_____	6.	Observe several burner ignitions, use power supply switch or SSU switch.
_____	7.	Comment on burner light off as far as flame light off shape, noise, and smooth and quick cut-off.
_____	8.	Observe blower startup.
_____	9.	Comment on belt tension, bearing noise, dirt in blower blades.

BLOWER MAINTENANCE

Check	Step	Procedure
_____	1.	Turn power off, test with meter to verify power off and install lockout tag and lock.
_____	2.	Remove wires from blower motor at an accessible location.
_____	3.	Remove screws or bolts securing blower assembly.
_____	4.	Remove blower assembly for cleaning and inspection.
_____	5.	Inspect belt for cracks and signs of wear.
_____	6.	Inspect pulleys for wear, grooving, and alignment.
_____	7.	Spin blower by hand, observe pulleys and blower wheel turn.
_____	8.	Inspect for pulley or blower wobble and alignment.
_____	9.	Listen for bearing noise, drag, or movement.
_____	10.	Inspect blower shaft for signs of wear.
_____	11.	Clean blower and motor with air pressure, brushes, and cleaning solution and water spray as required.
_____	12.	Oil motor and blower bearings as required.
_____	13.	Reassemble blower, taking care to check pulley alignment and correct belt tension. Check with instructor for correct blower operation.
_____	14.	Do not install blower until heat exchanger has been checked.

BURNER MAINTENANCE

Check	Step	Procedure
_____	1.	Remove and clean nozzle assembly.
_____	2.	Replace nozzle with a manufacturer recommended nozzle.
_____	3.	Place old nozzle in small zipper bag with new nozzle box. Write your name on the bag, date it, and leave it in the furnace burner compartment. The bag will keep it from smelling and next year or next week on any return visit, warranty or not, you will know what was put in and what you took out.
_____	4.	Set electrodes at manufacturer recommended setting. If not known, use the following default values. Use $1/8$ in apart, $1/2$ in up, and $1/16$ in forward of nozzle face.
_____	5.	Remove vent connector, barometric damper, and flue baffles or cleanout plugs.
_____	6.	Remove or swingout burner assembly.
_____	7.	Vacuum and brush all soot rust and solid particles from the fire side of the heat exchanger. Be careful not to damage combustion chamber.
_____	8.	Insert a light into combustion area.
_____	9.	If possible turn off lights in furnace room.
_____	10.	With the light in as far back into the heat exchanger as possible, inspect the heat exchanger from both the fan side and the plenum side for holes, light, or leaky gaskets.
_____	11.	Reinstall blower assembly after the heat exchanger check.
_____	12.	Reinstall nozzle assembly while burner is still out of or swung away from furnace.
_____	13.	Inspect positioning of nozzle and electrodes.
_____	14.	Reinstall or reposition burner on furnace.
_____	15.	Install an oil pressure gauge on pump supply line.
_____	16.	Remove lockout tag and lock, turn on burner, and check and record initial oil pressure. Pressure = _____ PSIG
_____	17.	Adjust to manufacturer recommended oil pressure. If no other pressure is listed, use 100 PSIG. Record final oil pressure. Pressure = _____ PSIG
_____	18.	Observe burner turn off due to flame failure.
_____	19.	Observe oil pressure gauge holding 85 lb quick cut-off.
_____	20.	Install original oil supply line to nozzle assembly.
_____	21.	Check and adjust fan control for correct fan and limit settings. (Limit = 200 max, recommended fan on = 135 and off = 90.) Record final settings. Limit = _____ On = _____ Off = _____
_____	22.	Obtain normal burner operation (fan has not been installed).
_____	23.	Observe burner operate with no fan until burner turns off (burner cycles on limit). This tests the limit check.
_____	24.	Reinstall blower assembly.
_____	25.	Reinstall all panels.
_____	26.	Check final operation on thermostat.
_____	27.	Demonstrate correct operation for customer. Be ready to demonstrate and explain any problems noted or corrected with equipment.

OIL BURNER COMBUSTION EFFICIENCY

STUDY MATERIAL
Chapter 11, Unit 2

LABORATORY NOTES

Combustion testing is used when performing a complete oil furnace preventive maintenance. Since all modern oil burners are high pressure gun type burners and these are power burners, you have complete authority over combustion air input to the burner. Oil burners can also be fired at an assortment of nozzle sizes calibrated in flow rate units of GPH (gallons per hour), which will change the combustion air requirement. We need to measure CO_2 with flue analysis to accurately set the combustion air. We typically look for a 10% CO_2 which will mean 50% excess air, to ensure complete combustion of the fuel.

Perfect combustion of a fuel is accomplished by having exactly the right amount of air mixed with exactly the right amount of fuel to burn the fuel. This mixture is dangerous to strive for since too little oxygen will waste fuel and produce carbon monoxide. There are no simple methods to test for unburned fuel in the vent gases. The ultimate CO_2 giving a perfect combustion mixture for the various fuels is listed below. Due to the difficulty of ensuring a complete mixing, it is best to burn any fuel with at least 50% excess air, this will ensure all the fuel is burned, with a minimum of stack loss.

Ultimate CO_2	Recommended CO_2		
(Perfect combustion)	25%	50%	75%
	Excess air	Excess air	Excess air
(Approximate values)			
Natural gas = 11.7 – 12.2	9%	8%	7%
Propane = 13.7	11.5%	9.5%	8%
No. 2 Oil = 14.7	12.5%	10.5%	9%

Ultimate CO_2	Recommended CO_2		
(Perfect combustion)	25%	50%	75%
	Excess air	Excess air	Excess air
(Approximate values)			
All fuels = 0%	5%	7%	9%

UNIT DATA

1. Shop ID # _____

2. Furnace make _____   Model # _____

3. Burner type (circle one):   Atmospheric   Induced Draft   Power

4. Nozzle GPH = _____

5. Estimated thermal efficiency = _____

PREPARATION FOR COMBUSTION TESTING

Check	Step	Procedure
_____	1.	Vent complete.
_____	2.	Fuel line complete and leak free.
_____	3.	No loose wires dangling unconnected.
_____	4.	All panels in place or adjacent to furnace.
_____	5.	Thermostat installed and operable.
_____	6.	Locate CO_2 test openings and temperature probe between furnace and barometric damper.

COMBUSTION TEST PROCEDURE

Check	Step	Procedure
_____	1.	Insert thermometer probe and CO_2 or O_2 probe into the breach, that is, between the furnace outlet and the barometric damper.
_____	2.	Turn on main power and thermostat to a call for heat.
_____	3.	Observe burner on and fan on.
_____	4.	Observe continuous burner operation until vent gas temperature tops out.
_____	5.	Record CO_2 (or O_2) and stack temperatures in the space provided below (Test 1 through Test 4 under readings).

CONTINUED COMBUSTION TEST

Check	Step	Procedure
_____	1.	Burner is warmed up with actual stack temperature topped out and reading a constant temperature.
_____	2.	Take initial readings and record in space provided.
_____	3.	Close primary air shutter until a lazy yellow flame is present. Test the yellow flame.
_____	4.	Open the air shutter to a maximum, test with excess air.

Check	Step	Procedure
_____	5.	Close the air shutter to obtain 8% CO_2.
_____	6.	Using the slide rule calculator with the No. 2 oil slide, look up and record the combustion efficiency.
_____	7.	Open the air shutter until the oil flame just begins to rumble. At this point, the flame is on the verge of blowing out due to too much air.
_____	8.	Measure and record reading in the space provided.
_____	9.	Close and open air shutter for the cleanest and quietist flame available.
_____	10.	Measure and fine tune to get 10% CO_2.

READINGS

	Test 1 (initial)	Test 2 (yellow)	Test 3 (excess air)	Test 4 (normal)
CO_2	_____	13%	9%	11%
Actual stack temperature	_____	_____	_____	_____
Net stack (gross, room)	_____	_____	_____	_____
Combustion efficiency	_____	_____	_____	_____

CONCLUSIONS

Notice that the yellow flame will produce a good combustion efficiency reading. This reading is actually invalid due to the unburned fuel going up the stack. All combustion testing equipment is accurate only when complete combustion of fuel is taking place. This is a common problem and source of confusion when using combustion testing equipment.

Question: Which flame is the most efficient? _____

Comments: _____

MEASURE OIL FURNACE THERMAL EFFICIENCY

STUDY MATERIAL
Chapter 11, Unit 2

LABORATORY NOTES
The purpose of this test is to measure thermal efficiency and to draw a correlation, if there is any, between the airflow through a furnace, and the thermal efficiency of the furnace. The furnace used for this test is a standard oil furnace using a high pressure gun type oil burner. The thermal efficiency of the furnace is calculated at three different airflows. This laboratory worksheet assumes an airflow hood is available to measure airflow through the furnace. Airflow can be measured on the return air opening to the furnace and avoid problems of some flow hoods being damaged due to the high temperature of the furnace discharge. When measuring airflow on the return air side, the airflow will go down as the burner comes on. This is due to expansion of air going through the furnace. It may be necessary to install a bracket to hold the airflow hood. The airflow could also be measured by the formula (velocity × area in units of ft^2) = airflow in units of CFM.

UNIT DATA

1. Shop ID # _____

2. Furnace make _____ Model # _____

3. Fuel _____ No. 2 Oil _____

4. Burner type (circle one): Power   X

5. Current nozzle GPH = _____

6. Calculate burner input by the following formula:

    GPH × 140,000 = Burner input

    _____ × 140,000 = _____

PRETEST OPERATION

Check	Step	Procedure
_____	1.	Obtain normal heating operation.
_____	2.	Observe system operation with a continuous burner operation.
_____	3.	Slow down blower speed or close off registers to obtain the lowest possible airflow through furnace and still maintain 100% burner operation. Burner must not cycle on limit.

TEST DATA

Check	Step	Procedure
_____	1.	Obtain fan only operation. Airflow hood may be damaged if exposed to temperatures above 140°F.
_____	2.	Measure furnace airflow in CFM on inlet and/or outlet using airflow hood or anemometer.
		Airflow = _____ CFM
_____	3.	Remove airflow hood from discharge of furnace when operating furnace in heating mode.
_____	4.	Operate furnace in normal heating mode.
_____	5.	Measure temperature rise through furnace. Temperature difference = _____
_____	6.	Calculate furnace Btu output by the following formula:

Air temperature rise × Airflow × 1.08 = Btu

_____ × _____ × 1.08 = _____

Check	Step	Procedure
_____	7.	Use Btu input calculated in Unit Data from actual nozzle.
_____	8.	Determine thermal efficiency of the furnace by the following formula:

BTU output / BTU input = % thermal efficiency

Test 1 (initial airflow): _____/ _____ = _____%

Check	Step	Procedure
_____	9.	Lower the furnace airflow and repeat the above test.

Test 2: _____/ _____ = _____%

Check	Step	Procedure
_____	10.	Raise the furnace airflow and repeat the above test.

Test 3. _____/_____ = _____%

Check	Step	Procedure
_____	11.	What correlation can be drawn between the airflow and the thermal efficiency of the furnace?

_____

Check	Step	Procedure
_____	12.	When the airflow increases, the thermal efficiency will (circle one):

Increase   Decrease   Remain the same

BASIC OIL FURNACE REPLACEMENT

STUDY MATERIAL
Chapter 11, Unit 2

LABORATORY NOTES

The simplest oil furnace replacement assumes that the old oil furnace has failed and is not worth repairing. This laboratory worksheet covers a replacement of the furnace only without any duct system changes or AC considerations. This is done under three reasons: first as a *give me your best price*; second, as a *get it done today emergency situation*; and third as a *warranty replacement*. The old furnace is disconnected, pulled out, and the new one is slid under the old duct connection and hooked back up.

REMOVAL OF EXISTING OIL FURNACE

Check	Step	Procedure
_____	1.	Turn off, lock out, and tag power to furnace at main electrical box.
_____	2.	Shut off oil supply valve at tank or burner.
_____	3.	Drain oil from line. Throw oil dry on any spilled oil.
_____	4.	Pull old oil line back out of the way; it may or may not be used again.
_____	5.	Remove electrical conduit from furnace to junction box, or main breaker terminal.
_____	6.	Disconnect vent pipe from furnace to chimney.
_____	7.	Clean up basement before installation of new furnace.

INSTALLATION OF NEW ASSEMBLED OIL FURNACE

Check	Step	Procedure
_____	1.	Remove furnace from carton.
_____	2.	Measure and cut in location of return air filter fitting. Install filter fitting.
_____	3.	Slide furnace under old return air and supply air plenum.
_____	4.	Install return air plenum fitting and pop rivet or screw to furnace.

Check	Step	Procedure
_____	5.	Install air filter.
_____	6.	Move furnace around for the simplest connection of supply air plenum, return air plenum, gas line, and electrical connections.
_____	7.	Measure and make connection to supply air plenum.
_____	8.	Measure and make connection to return air plenum.
_____	9.	Connect existing oil line to furnace. Add or replace pieces as required. Install bypass plug for two pipe installation.
_____	10.	Install new smoke pipe from furnace to chimney as recommended by manufacturer.
_____	11.	Install barometric damper in a location recommended by the furnace manufacturer.
_____	12.	Connect main power supply to furnace. Local code may require a licensed electrician for electrical connection.
_____	13.	Connect thermostat wire to furnace terminal strip per wiring diagram.
_____	14.	Check/replace thermostat as required.
_____	15.	Set thermostat anticipator at manufacturer recommended setting.
_____	16.	Check and record the nozzle that is installed.

Make _____   GPH _____   Angle _____   Pattern _____

Check	Step	Procedure
_____	17.	Turn oil on at valve and check for any leaks.
_____	18.	Snug nozzle, check and adjust electrodes, and install nozzle assembly.
_____	19.	Remove lockout tag and lock.
_____	20.	Turn on burner and bleed air from oil line (one pipe oil line only).
_____	21.	Observe oil burner establish a flame.
_____	22.	Adjust air shutter to obtain a clean flame.
_____	23.	Set air intake to 0 smoke.
_____	24.	Demonstrate normal operation to the customer.

TWO PIPE CONVERSION

STUDY MATERIAL
Chapter 11, Unit 3

LABORATORY NOTES

A one pipe oil system refers to the oil line from the tank to the burner being a single pipe, usually $3/8$ in OD copper. This system is recommended for no more than 2 ft vertical lift from the oil level in the tank to the burner pump. Any time the tank runs out of oil the air must be manually bled from the line at the pump. Either of these two service problems can be a reason to switch the system over to a two pipe system.

Two pipe conversion involves running a second line back to the tank and installing a bypass plug within the pump. This second line at the tank needs to go all the way to the bottom of the tank. If oil entered the top of the tank and fell to the bottom, you would hear the oil fall and splash whenever the oil level dropped below the line outlet. Frequently a tank duplex fitting is used on two pipe systems. This fitting is installed in the top of the tank and has two fittings in it that will allow a $3/8$ in line to be pushed through to the bottom of the tank and then pulled up 3–4 in. This is the correct position to install both the supply line and the return line. We don't want to pull any sludge off the bottom of the tank.

UNIT DATA

1. Shop ID# _____

2. Furnace make _____   Model # _____

3. Pump make _____   Model # _____

SYSTEM INSPECTION

Check	Step	Procedure
_____	1.	Identify current fuel oil piping system (circle one): One pipe Two pipe
_____	2.	Note lift from the lowest possible operating oil level to the burner pump inlet fitting.
		Lift = _____
_____	3.	Is there a bypass plug at the pump? Every new oil burner comes with a bypass plug in a cloth bag generally attached to the pump with string. Is it still there?
		(circle one): Yes No

Check	Step	Procedure
_____	4.	If yes, you have the plug you need. If no, you will need to get one.　　Pump model _____
_____	5.	Obtain a bypass plug as required.
_____	6.	Obtain sufficient length of $3/8$ in copper tubing and required brass fittings to make connection at pump.

CONVERT TO TWO PIPE SYSTEM

Check	Step	Procedure
_____	1.	Turn off fuel line at tank.
_____	2.	Turn off, lock out, and tag out main electrical power.
_____	3.	Disconnect existing one pipe line.
_____	4.	Loosen and remove two bolts holding pump in position.
_____	5.	Inspect pump fitting opening for return line to tank (pumps are generally labeled for openings and plug location).
_____	6.	Inspect pump for install location of bypass plug. This is generally a $1/8$ in or $1/16$ in female pipe thread inside the return line opening. Refer to manufacturer data for pump as required.
_____	7.	Hold pump at an upward angle. Using an Allen wrench of sufficient length, install and tighten bypass plug.
_____	8.	Install pipe to $3/8$ in flare fitting in pump.
_____	9.	Mount pump back on burner housing.
_____	10.	Run $3/8$ in copper line from pump outlet into top of oil tank and down into tank at 4 in from bottom of tank.
_____	11.	Snug fitting to hold line in place.
_____	12.	Open tank shutoff, remove lockout, and start burner. It will now self-bleed with bypass oil going back to tank.
_____	13.	Perform any other burner tuneup procedures required.

INSTALL REPLACEMENT OIL PUMP

STUDY MATERIAL
Chapter 11, Unit 3

LABORATORY NOTES
The pump on an oil burner is one of the most important parts of the burner assembly. Its job is to pull the oil from the tank and to deliver it to the combustion chamber at the recommended supply pressure, usually 100 PSIG or 140 PSIG. Always check burner nameplate to verify manufacturer recommended oil supply pressure. Minor adjustments can be made, but if it is worn out it must be replaced or rebuilt. Most pumps are exchanged at the service counter.

There are dozens of different pumps available. This is because there is more than one pump manufacturer, several major burner manufacturers, and different types of burner configurations. Not every wholesale house has every pump. Start by going to a dealer of the furnace type or burner type you have. Have the furnace and burner model and serial number and also date of install if a warranty is a possibility. In cold weather you may need to get it back on line the same day you take it out. In mild weather you may have a few days. Get the exact or equivalent replacement pump and the bypass plug for a two pipe system.

UNIT DATA

Furnace make _____ Model # _____ Serial # _____

Burner make _____ Model # _____ Serial # _____

Pump make _____ Model # _____ Serial # _____

CHECK EXISTING PUMP

Check	Step	Procedure
_____	1.	Identify piping system (circle one): One pipe Two pipe
_____	2.	What is the lift? Lift = _____
_____	3.	If lift is greater than 2 ft on a one pipe system we will need to convert to a two pipe system.
_____	4.	Shut burner off.
_____	5.	Install gauge on oil line to nozzle.
_____	6.	Start burner and record pump pressure. Bleed pump as required to obtain reading.
		Pressure = _____PSIG

209

Check	Step	Procedure
_____	7.	Adjust to 100 PSIG (or rated PSIG) if possible.
_____	8.	Turn burner off and record pump cut-off pressure. It should be close to 85 PSIG.
		Cut-off pressure = _____ PSIG
_____	9.	Install gauge on high pressure port of pump. For one example, the Mitco P115-2 Kwik-Check 2 pump test kit measures operating pressure under burn and quick cut-off.
_____	10.	Connect nozzle supply line to nozzle fitting on burner.
_____	11.	Measure and record pump pressure during a normal burn. Note that a weak pump can hold 100 lb on an initial test but fail to hold 100 lb during a normal burn mode. Pressure = _____ PSIG
_____	12.	Any pump failing to hold 85 PSIG cut-off or maintain 100 PSIG during the normal burn mode needs to be replaced.

REPLACE PUMP

Check	Step	Procedure
_____	1.	Obtain correct replacement pump. Pump # _____
_____	2.	Turn off fuel line.
_____	3.	Turn off, lock out, and tag out electrical power.
_____	4.	Disconnect existing oil line or lines.
_____	5.	Loosen and remove two bolts holding pump in position.
_____	6.	Change fittings to new pump. Change to two pipe system if required.
_____	7.	Bolt new pump in position.
_____	8.	Mount pump back on burner housing.
_____	9.	Open tank shutoff and start burner.
_____	10.	Perform any other burner tuneup procedures required.

CHECK/TEST A CAD CELL OIL BURNER PRIMARY CONTROL

STUDY MATERIAL
Chapter 11, Unit 3

LABORATORY NOTES

The cadmium sulfide cell (cad cell) of an oil burner primary control system proves the presence of an oil flame by observing the visible light from the flame. The cad cell's resistance is greatly reduced in the presence of light. The resistance must be high to enable the primary to initiate a trial for ignition, also called a dark start function. Once a flame is established, the light from the flame causes the cad cell's resistance to drop and the flame will continue. During the trial for ignition, a safety switch heater is energized that will open the safety switch contacts and lock out the burner if flame is not proved within the trial for ignition time, usually 30, 45, or 60 sec. This heater must cool off before the burner can be reset manually and start again. When flame is established and proved by the cad cell, the safety switch heater is deenergized and the contacts remain closed.

UNIT DATA

1. Shop ID# _____

2. Furnace make _____   Model # _____

3. Furnace type (circle one):   Upflow   Basement   Counterflow   Horizontal

4. Burner make _____   Model # _____

5. Nozzle type _____

MINI FURNACE STARTUP

Check	Step	Procedure
_____	1.	Turn on power supply switch or SSU switch on and thermostat to a call for heat.
_____	2.	Observe normal flame ignition.
_____	3.	Observe flame continue and furnace blower come on.
_____	4.	Turn thermostat down and observe burner turn off.
_____	5.	Observe furnace blower turn off within 3 min.

TEST CAD CELL

Check	Step	Procedure
_____	1.	Turn power supply switch or SSU switch off.
_____	2.	Remove screws from ignition transformer and swing open.
_____	3.	Locate and unplug cad cell from plug mount.
_____	4.	Inspect and wipe clean lens cover of cad cell.
_____	5.	Test and record ohms of cad cell when exposed to light. Resistance = _____ ohms
_____	6.	Test and record ohms of cad cell when covered. Resistance = _____ ohms
_____	7.	Insert cad cell into plug assembly.
_____	8.	Locate yellow wires from cad cell mount at primary control terminals F and F.
_____	9.	Remove cad cell wires from F and F and connect to ohmmeter.
_____	10.	Swing transformer slowly closed.
_____	11.	What happens to the resistance of the cad cell as the transformer closes? _____

CHECK PRIMARY CONTROL

Check	Step	Procedure
_____	1.	With ohmmeter still connected to cad cell and transformer closed, turn power supply switch or SSU switch on and thermostat to a call for heat.
_____	2.	Observe burner turn on and flame ignite.
_____	3.	Read and record ohms of cad cell exposed to a normal oil flame. Resistance = _____ ohms
_____	4.	Observe flame turn off and lock out after 30 sec.
_____	5.	Obtain a 1200 ohm resistor.
_____	6.	Connect one end of resistor to one F terminal.
_____	7.	Push reset button. Wait a minimum of 2 min cooldown time.
_____	8.	Observe burner start and flame ignite.
_____	9.	Connect second wire of 1200 ohm resistor to the other F terminal within the 30 sec trial for ignition time.
_____	10.	Observe flame continue for 2–5 min.
_____	11.	Obtain watch with second hand.
_____	12.	Remove one lead of resistor from F and begin timing.
_____	13.	Observe burner turn off and lock out.
_____	14.	Record run time after removing resistor. Time = _____
_____	15.	Repeat as required to get accurate lockout time. Reconnect resistor for continued run and remove to simulate a flame failure.
_____	16.	Monitor resistance of cad cell during normal flame.

Check	Step	Procedure
_____	17.	What causes the fluctuation of the ohms during a normal burn cycle of the flame?
		_____
_____	18.	Leave the 1200 ohm resistor connected to F and F.
_____	19.	Attempt to turn burner on with resistor connected.
_____	20.	Does burner turn on while connected to resistor?
_____	21.	Why or why not? _____
		_____

INSTALL REPLACEMENT OIL BURNER

STUDY MATERIAL
Chapter 11, Unit 1

LABORATORY NOTES
Many oil furnaces are constructed of heavy metal and are quite durable. Sometimes the burner just seems to be worn out, while the basic furnace is in good shape. In such cases the entire burner can be replaced. This has the advantage of a matched nozzle assembly and flame cone along with a new pump, motor, transformer, and primary control. In an oil heat section of the country this is a fairly common job. In this lab, it will be assumed that the customer has had problems with their old burner and has decided to get a new one installed. This is less expensive, faster, and easier that an entire new furnace or boiler. We will not need to check any of the old components before removing them.

UNIT DATA

1. Furnace make _____ Model # _____ Serial # _____

2. Existing burner make _____ Model # _____ Serial # _____

3. New burner make _____ Model # _____ Serial # _____

4. Length of blast tube required = _____

5. Identify type of piping system (circle one):   One pipe   Two pipe

6. New combustion chamber? (circle one):   Yes   No

7. New thermostat (circle one):   Yes   No

REMOVE OLD BURNER

Check	Step	Procedure
_____	1.	Turn off oil at tank or supply line.
_____	2.	Turn off, lock out, and tag out main power.
_____	3.	Disconnect fuel supply line and return lines.
_____	4.	Install ³/₈ in flare plug in both oil lines.
_____	5.	Carefully bend lines back out of the way. We will reuse the same lines if possible.

Check	Step	Procedure
_____	6.	Disconnect main power and thermostat wire from burner.
_____	7.	Remove mounting bolts from burner.
_____	8.	Remove mount plate from front of furnace or boiler.
_____	9.	Use a vacuum to clean any debris from combustion chamber area. Do not damage combustion chamber.
_____	10.	Inspect combustion chamber for any signs of cracks or deterioration. Replace as required.

INSTALL NEW BURNER

Check	Step	Procedure
_____	1.	Install new combustion chamber as required.
_____	2.	Hold burner mount plate in position and measure distance to combustion chamber. Distance = _____
_____	3.	Measure length of blast tube on new burner. Length = _____
_____	4.	Exchange blast tube on burner if length is not a match.
_____	5.	Check/install nozzle for correct GPH, angle, and pattern. Record data. GPH _____ Angle _____ Pattern _____
_____	6.	Check and adjust electrode position with booklet.
_____	7.	Bolt burner mounting plate to furnace or boiler.
_____	8.	Bolt new burner with blast tube onto burner plate.
_____	9.	Connect oil lines to burner. Install bypass plug for two pipe systems.
_____	10.	Replace oil filter in fuel supply line. If no oil filter is there, install a new one.
_____	11.	Open tank shutoff and bleed oil at pump to get oil through filter.
_____	12.	Turn power and burner on. Bleed pump as required.
_____	13.	Perform any other burner tuneup required.

ELECTRIC FURNACE LABORATORIES

ELECTRIC FURNACE STARTUP

STUDY MATERIAL
Chapter 12, Unit 1

LABORATORY NOTES
A typical electric furnace will have three stages of electric heat, usually 3 or 5 KWH each. One KWH would be 3400 Btu/hr and pull 4.16 amps at 240 V. A 3 KWH heater would be three times that or 12.5 and a 5 KWH heater would be 20.8 amps. Three 5 KW heaters would pull over 60 amps, too much to just turn on. Power companies and some codes require electric furnaces to be equipped with a sequencer that is a time delay device, and multiple contactors. This is installed in the furnace and not part of the thermostat. Even with a single-stage heat-only furnace the electric heater elements come on one at a time, spaced apart by a few seconds at least. Supply air temperatures of lower than 120°F can feel rather cool and airflow should be reduced to keep the air temperature to a comfortable level.

UNIT DATA

1. Furnace make _____   Model # _____

2. Electrical data: Voltage = _____   Phase = _____   Amps = _____   KW = _____

3. Blower type (circle one):   Direct drive   Belt drive

4. Blower motor speeds (circle one):   Single speed   2   3   4

5. System type (circle one):   Heat only   Heat pump and electric backup

6. Electric heaters: 1st = _____KW   2nd = _____KW   3rd = _____KW   Total KW= _____

PRESTART CHECKS

Check	Step	Procedure
_____	1.	Check all electrical connections for tightness.
_____	2.	Spin all fans to be sure they are loose and spin freely.
_____	3.	Airflow passages are unobstructed.
_____	4.	All doors and panels available and in place.
_____	5.	Thermostat installed and operating correctly.

START UP AND CHECK OPERATION

Check	Step	Procedure
_____	1.	Turn power off and set thermostat to below room temperature.
_____	2.	Measure voltage at plug to be used. Voltage = _____
_____	3.	Does voltage match nameplate requirement? (circle one): Yes No
_____	4.	Plug unit in and turn power on.
_____	5.	Obtain fan only operation, use thermostat fan on, push button of fan limit, or consult instructor.
_____	6.	Turn fan off and thermostat to heat.
_____	7.	Adjust temperature setting to 10°F above room temperature.
_____	8.	Observe fan on and heater banks come on.
_____	9.	Measure and record amperage as sequencer brings on electric banks.
		1st = _____ 2nd = _____ 3rd = _____
_____	10.	Turn thermostat down and observe heaters off.
_____	11.	Observe blower off within three min of heaters off.

READINGS AND MEASUREMENTS

Check	Step	Procedure
_____	1.	Obtain normal heating operation.
_____	2.	Measure and record temperatures.
		Discharge air temperature = _____ Room air temperature = _____
_____	3.	Calculate temperature rise by the following formula:

Discharge air temperature − Room air temperature = Temperature rise = _____

_____ − _____ = _____

CALCULATE AIRFLOW BY TEMPERATURE RISE

STUDY MATERIAL

Chapter 12, Unit 2

LABORATORY NOTES

Electric heaters are exactly 100% efficient and make the heat produced very accurate. The heaters are located within the duct or furnace and there is no lost heat going up the chimney. We can calculate the airflow by the temperature rise method very accurately. The trick is to accurately measure the voltage, amperage, and temperatures. We also must separate the electric heater amperage from the fan amperage.

UNIT DATA

1. Furnace make _____ Model # _____

2. Electrical data: Voltage = _____ Phase = _____ Amps = _____ KW = _____

3. Blower type (circle one): Direct drive Belt drive

4. Blower motor speeds (circle one): Single speed 2 3 4

5. System type (circle one): Heat only Heat pump and electric backup

6. Electric heaters: 1st = _____KW 2nd = _____KW 3rd = _____KW Total KW = _____

PRESTART CHECKS

Check	Step	Procedure
_____	1.	Spin all fans to be sure they are loose and spin freely.
_____	2.	All doors and panels available and in place.
_____	3.	Thermostat installed and operating correctly.
_____	4.	Locate and count each heater element contractor. How many of what KW are there?
		_____ of _____ KWH

START UP AND CHECK OPERATION

Check	Step	Procedure
_____	1.	Turn power on and obtain a normal heating mode.
_____	2.	Observe all heater elements operating.
_____	3.	Measure and record amperage as sequencer brings on electric banks.
		1st = _____ 2nd = _____ 3rd = _____
_____	4.	Turn thermostat down and observe heaters off.
_____	5.	Observe blower off within two min of burner off.

READINGS AND MEASUREMENTS

Check	Step	Procedure
_____	1.	Obtain normal heating operation.
_____	2.	Measure and record temperatures.
		Discharge air temperature = _____ Room air temperature = _____
_____	3.	Calculate temperature rise by the following formula:

Discharge air temperature – Room air temperature = Temperature rise = _____

Check	Step	Procedure
_____	4.	Read total amperage of electric heaters only. Amps = _____
_____	5.	Calculate Btu/hr by the following formula:

Voltage × Amps × 3.414 Btu/hr = _____

_____ × _____ × _____ = _____

Check	Step	Procedure
_____	6.	Use the following formula to calculate airflow by the temperature rise method.

$$\text{Airflow} = \frac{\text{BTU output from step \#5}}{1.08 \times \text{Temperature rise from step \#3}} = \underline{\hspace{1cm}} \text{ CFM}$$

Check	Step	Procedure
_____	7.	Measure fan motor amperage. Amps = _____
_____	8.	Read rated fan motor amperage from nameplate or motor. Amps = _____
_____	9.	Blower motor amperage should be lower than the blower motor rating.
_____	10.	Call your instructor over to demonstrate furnace operation.

REFRIGERATION LABORATORIES

BASIC REFRIGERATION SYSTEM STARTUP

STUDY MATERIAL

Chapter 5, Unit 1

LABORATORY NOTES

This basic startup assumes the system had been working in the past and is still expected to be operable but hasn't been used for several months. Similar to an ice cream parlor that has been closed for the winter and will open for the summer, or a school that has been closed for the summer. It is your job to perform the pre-start visual inspection, install the gauges, check system idle pressures, and to operate the system in a normal mode, making a judgment as to current system operation condition.

UNIT DATA

1. Shop ID # _____

2. Unit description _____

3. System application and refrigerant type _____

SYSTEM INSPECTION, PRESTART CHECKS

Check	Step	Procedure
_____	1.	System condenser type (circle one): Evaporative Water Air
_____	2.	Compressor type (circle one): Hermetic Open Semihermetic
_____	3.	Evaporator type (circle one): Fan coil Plate type Other _____
_____	4.	Metering device (circle one): TXV Capillary tube Automatic Other _____
_____	5.	Line sizes. Liquid = _____ Suction = _____ Hot gas = _____
_____	6.	All lines are complete, connections tight, and apparently free from leaks? (circle one): Yes No
_____	7.	Predict the normal operating pressures using the ambient temperature, refrigerant type, and the system application. High side pressure = _____ Low side pressure = _____

SYSTEM OPERATION

Check	Step	Procedure
_____	1.	Install gauges and record the system idle pressures.
		High side pressure = _____ Low side pressure = _____
_____	2.	Call your instructor over if the system idle pressures are less than saturated condition for refrigerant type.
_____	3.	System verified ready to perform a startup.
_____	4.	Turn system on.
_____	5.	Observe high side pressure go up and low side pressure go down.
_____	6.	Allow 5 min of operation for system to obtain a normal operation.
_____	7.	Record actual stabilized system pressures.
		High side pressure = _____ Low side pressure = _____
_____	8.	Approximate temperature (by finger touch only) check:
		_____ a. Liquid line is warm.
		_____ b. Suction line is cool.
		_____ c. Refrigerated space is cooling off.
		_____ d. Condenser discharge air is warm.

SYSTEM OPERATING CONCLUSIONS

Check	Step	Procedure
_____	1.	Is system fully charged and operating normal? (circle one): Yes No
_____	2.	What if any improvements could be made to system? _____
		_____
_____	3.	Call your instructor over for evaluation of operation to observe and removal of gauges.
_____	4.	Gauges removed, salvaging all possible refrigerant.
_____	5.	Turn system off.
_____	6.	Install caps on all stems and service ports.

REFRIGERANT RECOVERY

STUDY MATERIAL
Chapter 6, Unit 3

LABORATORY NOTES
This lab is intended to help students practice the removal of refrigerant from a system. This is a job that must be done on any and all systems prior to any refrigerant side repairs or dismantling. In the labs that follow, recovery will be listed and required on all jobs performing any major refrigerant service. This first time a student is performing a refrigerant recovery it is a job by itself. In future labs and jobs, recovery will be in addition to the bigger job of making the repairs, instead of a job by itself.

UNIT DATA

1.   Shop ID # _____

2.   Unit description _____

3.   System refrigerant.   Type _____   Amount _____

4.   Factory test pressure.   High side pressure = _____   Low side pressure = _____

5.   Determine target system vacuum level from Table 6-3-1 of *Refrigeration and Air Conditioning, 4E*, on page 327.

6.   Vacuum target pressure (circle one):   0 in Hg   4 in Hg   10 in Hg   15 in Hg.

RECOVER EXISTING REFRIGERANT, BASIC METHOD

Check	Step	Procedure
_____	1.	Install gauges and record system idle pressures.
		High side pressure = _____ Low side pressure = _____
_____	2.	Obtain recovery station, accurate scale, target cylinder partially filled with system refrigerant type, and one extra hose.
_____	3.	Weigh the target tank and check refrigerant type. Weight = _____
_____	4.	Read and record tare weight listed on tank. Weight = _____

Check	Step	Procedure
	5.	Subtract #4 from #3 to calculate the weight of refrigerant in the tank.
		Weight of refrigerant = _____
	6.	Connect gauge manifold center hose to the inlet of the recovery station (vapor side valve).
	7.	Connect extra hose from recovery station to the vapor side of the target tank.
	8.	Open low side gauge handle and purge center hose at recovery station as required.
	9.	Open recovery station outlet valve and purge hose at target tank.
	10.	Plug in recovery station and turn on pump.
	11.	Observe recovery outlet gauge going up.
	12.	Open target tank inlet valve.
	13.	Observe low side gauge manifold pressure dropping.
	14.	Hear and feel refrigerant entering target tank.
	15.	Observe target tank increase in weight as refrigerant enters tank.
	16.	Do not exceed listed max weight of tank.
	17.	Open gauge manifold high side pressure gauge handle to continue recovery from high side of system.
	18.	Continue recovery process until desired system pressure is reached.
	19.	Close both gauge handles, turn off recovery station, and observe system pressure.
	20.	Turn on recovery station to bring system pressure back down as required.

LEAK TEST AND EVACUATION

STUDY MATERIAL
Chapter 6, Units 3 & 4

LABORATORY NOTES

This lab is intended to allow student practice on any of the three basic leak testing and evacuation procedures commonly used in the field: timed or 1 hr method, triple evacuation (triple vac), or a micron evacuation. Since it is not the intention to recharge the system, an unrepaired leak could be set up and left for students to find. On such a system it would be impossible to pass a micron evacuation or any vacuum pressure drop test.

When evacuating any system using Schrader valves, the cores are sometimes removed. The purpose is to get a better quality evacuation faster. Some service technicians instructors may not want to bother with this practice. If the cores are removed, they must be put back in. Some people feel that the time and effort required to put the cores back in and the danger of getting contamination back into the system in the process make it an ineffective practice.

I have noted the time at which the cores should be removed and put back in but this is optional. Consult your instructor before removing Schrader cores and put them back in if you take them out.

UNIT DATA

1. Shop ID # _____

2. Unit description _____

3. System application and refrigerant type _____

4. Factory test pressure.   High side pressure = _____   Low side pressure = _____

RECOVER EXISTING REFRIGERANT

Refer to Laboratory Worksheet R-2 for the refrigerant recovery procedure.

LEAK TEST PROCEDURE

Check	Step	Procedure
_____	1.	Install gauges and record system idle pressures.
		High side pressure = _____ Low side pressure = _____
_____	2.	Inspect system for signs of oil, soap, nonoriginal soldering, metal deterioration, or any signs of leak problems.
_____	3.	Obtain a halide tester, electronic tester, and soap bubbles tester.
_____	4.	Verify correct operation of both halide and electronic testers with a reference leak.
_____	5.	Use existing charge (if any) for first stage of leak testing. If there is no charge, proceed to the second stage.
_____	6.	Recover existing charge as required. (Check with instructor.)
_____	7.	Pressurize system to 25 PSIG R-22 and boost to 100 PSIG with the nitrogen, helium, or CO_2 for a legal leak testing gas. Obtain a demonstration on the use of the nitrogen cylinder as required.
_____	8.	Leak test entire system with both halide and electronic testers.
_____	9.	Use the soap bubbles test to pinpoint any suspected leaks.
_____	10.	Note all leak locations and consult your instructor before making any repairs. Remember this saying: *Good leaks are hard to find*.
_____	11.	System certified leak free and ready for evacuation.

PART 1: EVACUATION METHOD 1: TIMED METHOD, ONE HOUR

The assumption is made that 1 hr of evacuation will remove system contamination without actually measuring the degree of remaining contamination. This method is still used by many service technicians. The time is sometimes varied to fit time available. An evacuation of only 1/2 hr might be good enough if the system is known to be clean. One hour or longer would be even better, no matter what the system problems. In moisture removal processes overnight evacuations are common.

Check	Step	Procedure
_____	1.	Bleed leaktesting mixture to 0 PSIG.
_____	2.	[Optional] Remove Schrader cores (as required), or move service valve stems to the intermediate position.
_____	3.	Connect vacuum pump to the center hose of manifold.
_____	4.	Close both gauge handles.
_____	5.	Turn on vacuum pump.
_____	6.	Open both gauge handles slowly. This will prevent excess vapor flow through vacuum pump from blowing oil out the discharge. This is more important on larger systems that may have residual pressure remaining in them.
_____	7.	Open gauge handles completely.
_____	8.	Observe low side gauge going down and approaching 30 in Hg vacuum.
_____	9.	Record time system reaches 30 in Hg. Time = _____
_____	10.	This is the beginning of your evacuation time. Call your instructor over to verify.

PART 2: VACUUM PRESSURE DROP TEST TIMED

The timed vacuum pressure drop test is typically done at the conclusion of a timed evacuation or a triple vac and used as an additional leak testing procedure. If the system holds 30 in Hg overnight or over a weekend, there can be no leaks. This is a valid procedure and should be done as time permits at the discretion of the service technician and company policy.

Check	Step	Procedure
_____	1.	Observe evacuation continuing at 30 in Hg and record time. _____
_____	2.	Close both gauge handles with vacuum pump still operating.
_____	3.	Close blank off valve on vacuum pump.
_____	4.	Remove evacuation hose from pump connect to blank spud on gauge manifold.
_____	5.	Observe that the system is still at 30 in Hg and record time. _____
_____	6.	Put vacuum pump away.
_____	7.	Call your instructor over for inspection.

PART 3: TRIPLE EVACUATION

The triple evacuation (triple vac) procedure consists of three consecutive evacuations spaced by two dilutions of a dry gas. Nitrogen is preferred but helium and CO_2 can also be used. The clean dry gas will act as a carrier, mixing with system contamination (air and water) and carrying it out during the next evacuation. It is a time consuming procedure but effective in obtaining a clean dry system.

Check	Step	Procedure
_____	1.	Bleed system to 0 PSIG before connecting vacuum pump.
_____	2.	[Optional] Remove Schrader cores (as required), or move service valve stems to the intermediate position.
_____	3.	Begin evacuation and observe pressure dropping. Observe evacuation reach 30 in Hg.
_____	4.	At the conclusion of the first evacuation follow the following steps to obtain a dilution pressure.

    _____ a. Close both gauge handles at the manifold.

    _____ b. Turn off vacuum pump.

    _____ c. Disconnect charging hose from pump.

    _____ d. Install charging hose on the dry gas cylinder pressure regulator output.

    _____ e. Open cylinder and increase pressure output to 50 lb.

    _____ f. Purge air from the center charging hose at the manifold.

    _____ g. Open both gauge handles slowly and obtain the desired dilution pressure.

Check	Step	Procedure
_____	5.	Record times for evacuation and dilution time and pressure. Evacuation and dilution pressure can be varied at the discretion of your instructor. Check off each item as you come to it below.

	Evacuation time	Dilution pressure	Dilution time
1st	___ 30 min ____	___ 10 lb ____	___ 10 min ____
2nd	___ 30 min ____	___ 10 lb ____	___ 10 min ____
3rd	___ 30 min ____		

PART 4: MICRON EVACUATION

The evacuation pump must be equipped with a blank off valve at the pump. The purpose of the valve is to isolate the pump from the system and leave the micron gauge exposed to the system. If the system maintains under 500 microns for a 10 min vacuum pressure drop test, the system is considered clean, dry, and leak free. If 800 to 1200 microns is maintained, moisture is present in the oil. A leak will cause the micron gauge to rise steadily. A large leak will cause a rapid rise while a small leak will cause a slow rise. Free water in the system will cause a rise to about 20,000 microns.

CHECK VACUUM PUMP

Check	Step	Procedure
_____	1.	Obtain good quality vacuum pump and micron gauge.
_____	2.	Change oil in vacuum pump as required.
_____	3.	Connect vacuum pump to micron gauge only.
_____	4.	Check operation of pump and gauge by pulling a vacuum on sensor.
_____	5.	If vacuum reaches 200 microns or lower then the vacuum pump is OK.

MICRON EVACUATION

Check	Step	Procedure
_____	1.	Bleed leak testing mixture from system to 0 PSIG.
_____	2.	[Optional] Remove Schrader cores (as required), or move service valve stems to the intermediate position.
_____	3.	Install micron gauge in vertical position preferably at vacuum pump and with a separate valve.
_____	4.	Begin evacuation and observe pressure dropping.
_____	5.	Allow 15 min of evacuation time before exposing vacuum gauge sensor. This will keep oil from depositing on pressure sensor.
_____	6.	Open valve to micron gauge sensor and observe pressure dropping.
_____	7.	Record lowest pressure in microns reached. Pressure = _____ microns
_____	8.	Close both gauge handles to determine quality of evacuation microns pump is now pulling. Pressure = _____ microns
_____	9.	Open gauge handles and continue evacuation of system until system vacuum bottoms out. Pressure = _____ microns
_____	10.	When system pressure reaches approximately 300 microns you are ready to perform a vacuum pressure drop test.

MICRON GAUGE, VACUUM PRESSURE DROP TEST

Check	Step	Procedure

_____ 1. Record pressure in microns while vacuum is in progress.   Pressure = _____ microns

_____ 2. Close blank off valve leaving system exposed to micron gauge.

_____ 3. Turn off vacuum pump.

_____ 4. Observe micron pressure gauge seek true level of pressure in system.

_____ 5. Record pressure after 10 min.   Pressure = _____ microns

_____ 6. Use the following criteria to determine system condition.

     _____ a. A clean dry system will hold 500 microns or less.

     _____ b. A system with some moisture or contamination mixed with the system oil will level out at somewhere between 1000 and 1500 microns.

     _____ c. A system with a leak will have first show a continuous pressure rise on the micron guage and then the low pressure gauge manifold will begin to go up.

     _____ d. A system with free water (water not mixed with oil) will climb off the scale on the micron gauge while the compound gauge stays at 30 in Hg.

_____ 7. System passes pressure drop test if it does not climb above 500 microns within 10 minutes.

_____ 8. Systems not passing can be cleaned up by the following steps.

     _____ a. Repeat leak testing.

     _____ b. Replace filter drier.

     _____ c. Change compressor oil.

     _____ d. Combine the triple vac with the micron method.

     _____ e. Change oil in the vacuum pump.

     _____ f. Retest the vacuum pump, gauge manifold, and hoses to verify that your equipment will hold 500 microns for the 10 min vacuum pressure drop test.

     _____ g. Replace the hoses or the manifold as required.

_____ 9. System certified to have maintained under 500 microns for 10 min.

DOMESTIC CAPILLARY TUBE SYSTEM RECHARGE

STUDY MATERIAL
Chapter 6, Units 3 & 4

LABORATORY NOTES
This laboratory worksheet will follow the method for a one hour evacuation and charge by weight. A domestic capillary tube system is a critical charge system, meaning it takes an exact amount of refrigerant, generally weighed in to the ounce. The charge will be listed on the system nameplate. If the system has been altered this listed charge will be wrong and the system will have to be charged by the superheat method. Even when the exact weight is listed and correct, the charge should still be checked by the superheat method after sufficient run time for stabilization.

UNIT DATA

1. Shop ID # _____

2. Unit make _____   Model # _____

3. Unit description _____

SYSTEM INSPECTION

1. Condenser type (circle one):   Static   Forced air   In the wall

2. Evaporator type (circle one):   Fan coil   Cold call   Cold shelf

3. System Type (circle one):   Auto defrost   Air spillage   Refrigerant spillage

4. Cabinet Type (circle one):   Chest   Side by side   Top freezer   Bottom freezer

LEAK TEST

Check	Step	Procedure
_____	1.	Pull unit plug (system must be off during the leak testing process).
_____	2.	Locate access valves on high and low side of system.
_____	3.	Obtain a gauge manifold with quick seal ends or manual seal ends on all three hoses.

Check	Step	Procedure
	4.	Install gauges and record system idle pressures.
		High side pressure = _____ Low side pressure = _____
	5.	Recover existing refrigerant pressure required by Federal guidelines (10 in Hg). Record recovery pressure. Pressure = _____
	6.	Pressurize system to 25 PSIG R-22 and boost to 100 PSIG with the nitrogen, helium, or CO_2 for a legal leak testing gas. Obtain a demonstration on the use of the nitrogen cylinder as required.
	7.	Leak test with halide, soap bubbles, and electronic testers.
	8.	Record location of any leaks found.

EVACUATION (ONE HOUR METHOD)

Check	Step	Procedure
	1.	Put away all leak testing equipment.
	2.	Bleed system pressure to 0 PSIG through the center hose, catch any oil with a rag. [Optional] Remove Schrader valve cores manually if desired.
	3.	Obtain vacuum pump and extension cord.
	4.	Check oil level in vacuum pump.
	5.	Connect center manifold hose to vacuum pump.
	6.	Open both gauge handles on manifold.
	7.	Turn on vacuum pump.
	8.	Observe compound gauge drop to 30 in Hg.
	9.	Allow 1 hr of evacuation time at 30 in Hg to ensure a quality evacuation.
	10.	Record total evacuation time. Time = _____
	11.	Close gauge handles on manifold.
	12.	Turn vacuum pump off and put away.
	13.	Connect center hose to manifold.
	14.	Perform vacuum pressure drop test by maintaining 30 in Hg for 48 hr (as desired or time permits).
	15.	Observe system at 30 in Hg after 48 hr.

RECHARGE SYSTEM

Check	Step	Procedure
	1.	Read system nameplate for refrigerant type and amount. _____ oz of R-_____
	2.	Obtain an accurate scale (electronic recommended).
	3.	Obtain a refrigerant cylinder with sufficient refrigerant to fill system.
	4.	Weigh and record total weight of cylinder and refrigerant. Weight = _____

Check	Step	Procedure
_____	5.	Connect center hose to refrigerant cylinder. Use liquid port when charging with any R-400 series refrigerant or recovered refrigerant.
_____	6.	Make sure gauge manifold valves are closed.
_____	7.	Open cylinder valve to put pressure on center charging hose.
_____	8.	[Optional] If the Schrader valve cores have been removed, this is the time to replace them. Fill system to 2 PSIG and manually install cores (do not overtighten).
_____	9.	Open high (red) pressure gauge handle putting refrigerant into high side of system only.
_____	10.	Observe pressure rise on low pressure gauge. You now know that you have normal flow of refrigerant from high to low side of system.
_____	11.	Add refrigerant until correct weight is installed. Refrigerant can be put into low side also once normal flow have been verified.
_____	12.	Close both high and low pressure gauge handles and tank valve.
_____	13.	Turn system on.
_____	14.	Observe high side pressure go up and low side pressure go down.
_____	15.	Drain center hose into low side by opening low side gauge handle. Observe low side pressure go up and then down.
_____	16.	Close low side gauge handle.
_____	17.	Allow sufficient run time for system stabilization.

FINAL TEMPERATURE PRESSURE CHECK

Check	Step	Procedure
_____	1.	Measure and record the following.
		High side pressure = _____ Low side pressure = _____
_____	2.	Suction line temperature = _____
_____	3.	Liquid line temperature = _____
		Comment on system operation. _____

LEAK TEST, EVACUATE, AND RECHARGE; TXV TRAINER VERSION

STUDY MATERIAL
Chapter 6, Unit 3

LABORATORY NOTES

This laboratory will use the three stage nitrogen boost, triple evacuation (triple vac), charge by sight glass method. This worksheet will cover the job that is the most common of all refrigeration jobs, the leak test, evacuate, and recharge. It's a job that must be done to every refrigeration system whenever it's torn apart, changed, or fabricated. The evacuation method used in this worksheet is the triple vac method, which is three consecutive evacuations separated by two dilutions of 10 min in length of 10 lb positive pressure of an inert gas. We will perform this job on a thermostatic expansion valve trainer station and use the sight glass method of charging. When the system is charged it will have close to normal operating pressures, a warm liquid line, a cool suction line and a clear sight glass filled with liquid refrigerant.

UNIT DATA

1. Compressor model # _____ Trainer # _____ (inspect equipment now but look up compressor application BTU and rating point during evacuation process)

2. Temperature range _____ Refrigerant type _____

3. Compressor model # _____

4. Bill of material (BOM) # (if available) _____

5. Compressor BTU or CFH (if available) _____ Rating point _____

LEAK TEST DATA

Check	Step	Procedure
_____	1.	Record factory rated maximum test pressures.
		High side pressure = _____ Low side pressure = _____
_____	2.	Install gauges and record system idle pressures.
		High side pressure = _____ Low side pressure = _____

Check	Step	Procedure
_____	3.	Recover existing refrigerant pressure required by Federal guidelines (10 in Hg)
		Record recovery pressure: _____
_____	4.	Pressurize system to 25 PSIG R-22 and boost to 100 PSIG with the nitrogen, helium or CO_2 for a legal leak testing gas. (Obtain demonstration on the use of the nitrogen cylinder as required.)
_____	5.	For the third leak test, the leak test pressure can be boosted to higher pressure for additional leak testing and location of hard to find leaks. However, do not exceed factory maximum test pressure listed on the nameplate.
_____	6.	Use the following test methods: Soap bubbles, Halide, Electronic
_____	7.	Check at the following generic locations. TXV outlet, all other flare connections, field soldered connections, control access connections, service valves and ports, gauge hose connections, other (list)
_____	8.	List locations of leaks found and repaired. Repeat as required.
_____	9.	System certified leak free and ready for evacuation.

EVACUATION (TRIPLE EVACUATION PROCEDURE)

Check	Step	Procedure
_____	1.	Bleed leak testing pressure to 0 PSIG.
_____	2.	Install vacuum pump to center hose.
_____	3.	Open both gauge handles.
_____	4.	Check system hose connections for system access.
_____	5.	Turn on vacuum pump.
_____	6.	Observe compound gauge reaching 30 in Hg vacuum.
_____	7.	Evacuate for 30 min with the gauge on 30 in Hg.
_____	8.	Perform the first dilution with the system valves closed off and 10 lb of nitrogen introduced into the system.
		_____ a. Close gauge handles.
		_____ b. Turn off pump.
		_____ c. Move charging hose from pump to cylinder.
		_____ d. Open nitrogen cylinder valve pressurizing hose.
		_____ e. Purge charging hose at manifold for 1 sec.
		_____ f. Open both gauge handles until the gauge reads 10#
		_____ g. Leave the clean dry gas in the system for 10 min.
		_____ h. Bleed the system to 0 PSIG and begin the second evacuation.
_____	9.	Evacuate for 30 min at 30 in Hg.

Check	Step	Procedure
_____	10.	Perform the second dilution for 10 min. Follow steps a–h above.
_____	11.	Begin the last evacuation for a 30 min hold.
_____	12.	With vacuum pump operating, close both gauge handles turn vacuum pump off, move center hose to manifold.
_____	13.	Observe 30 in Hg on the low side (blue) gauge.
_____	14.	Perform a vacuum pressure drop test as desired and time permits. Circle the time of the test you performed: 1 hr 12 hr 24 hr 48 hr

CHARGING

Check	Step	Procedure
_____	1.	Following an overnight vacuum pressure drop test, some technicians like to put the vacuum pump on and reevacuate for 15 min before charging.
_____	2.	Obtain refrigerant cylinder with sufficient amount of correct refrigerant for recharge and accurate scale.
_____	3.	Record weight of refrigerant cylinder and refrigerant.
_____	4.	Follow the same procedure as # 8 from above to obtain a properly valved off evacuated system.
_____	5.	Connect refrigerant charging hose to refrigerant cylinder. Use liquid valve when charging any 400 series or any recovered refrigerant.
_____	6.	Open high pressure (red) gauge handles to fill system high pressure side first.
_____	7.	Observe refrigerant flow into low pressure side by low pressure gauge rising.
_____	8.	Close both gauge handles.
_____	9.	Turn system on. Observe high side pressure increase and low side pressure drop.
_____	10.	Add refrigerant into the low side until the high side pressure reaches 120 lb.
_____	11.	Allow the system to run. When you are getting close to a full charge, observe some liquid in the sight glass.
_____	12.	Add refrigerant slowly until you can match these five criteria as closely as possible. An improperly adjusted TXV will prevent you from obtaining these five matching conditions.

    _____ a. The high side pressure corresponds with the predicted high side pressure.

    _____ b. The low side pressure corresponds within the predicted low side pressure.

    _____ c. The sight glass just clears, meaning liquid.

    _____ d. The suction line is cool to the touch.

    _____ e. The liquid line is warm to the touch.

CALIBRATE A LOW PRESSURE CUT-OUT FOR SAFETY CUT-OUT OR CYCLE DEFROST

STUDY MATERIAL
Chapter 8, Unit 2

LABORATORY NOTES

A system using a low pressure cut-out will use it for either low pressure safety or for maintaining space temperature. In the low pressure safety function the low pressure cut-out will be calibrated to turn off the system on a loss of refrigerant cut-out at 0 and cut-in at 10. Sometimes the terms *high event* and *low event* are used for cut-in and cut-out. A pressure cycle defrost can be performed without the use of auxiliary heat when the evaporator space is 36°F or higher. Ice does melt at 32°F but the length of time required to melt ice with 34°F air makes the off cycle too long. For this reason, the space temperature rises to 36°F or higher during the off cycle. The absence of ice on the evaporator is measured by the temperature in the coil. When using a fan coil evaporator this can be done by a low pressure control, at 36 lb R-12 the evaporator is at 39°F. This fixed cut-in or high event of 36 lb ensures a defrost every off cycle and the variable cut-out or low event is set at a low enough pressure (14 lb for R-12) to ensure proper temperature and normal on/off cycles. The low pressure function also ensures that the low pressure side of the system will not go below 0 PSIG to ensure that air will not be drawn into the system. The pressure cycle defrost system uses one low pressure control to control three things: defrost, space temperature, and low pressure safety. The pressures must be calibrated with a gauge as the numbers on the control are not accurate enough. The idea is to obtain the pressure desired and adjust the control to turn on or off at that pressure. There are two adjustment screws; the range screw moves both the high event and the low event while the differential screw moves only the low event.

This job can be performed on any medium temperature trainer, walk in or reach in cooler equipped with a low pressure cut-out and a TXV. It must be remembered that the numbers on the low pressure cut-out control are not to be trusted. The pressure is to be set with a gauge that is of known accuracy. The function of calibrating is to observe the control change position at the desired pressure on your gauge. It is normal for a control to vary one pound from cycle to cycle but too much means an inaccurate control and is reason for condemning the control. If you find a control that fails to turn on or off at a regular pressure, demonstrate this to your instructor and replace the control.

UNIT DATA

1. Shop ID # _____

2. Refrigerant type   R-_____

3. Calculate pressure cycle defrost pressure settings any refrigerant at the following points:

    a. Cut-in. The pressure corresponding to 39°F.   Cut-in = _____

    b. Cut out. The pressure corresponding to 10°F.   Cut-out = _____

4. Low pressure cut-out make and pressure range _____

5. TXV inspection. Is system a TXV system? (circle one):   Yes   No

---

CALIBRATE FOR LOW PRESSURE SAFETY CUT-OUT

Check	Step	Procedure
_____	1.	Install gauges and observe system normal operation.
_____	2.	Locate low pressure cut-out and identify the two adjustment screws.
_____	3.	Remove and save the field adjustment knob from the range screw if there is one. Reinstall it after completion.
_____	4.	Identify the range screw and the differential screw. The range screw must be set to fix the high event before the low event can be set correctly.
_____	5.	Using the range screw, adjust the high event or cut-in at 10 lb pressure.
_____	6.	Using the differential screw, adjust the low event or cut-out to 10 lb pressure. The cut-in will be at 10 lb and the cut-out will be at 0 lb.
_____	7.	Front seat the receiver service valve.
_____	8.	Observe the system begin pumping down.
_____	9.	The compressor should turn off at 0 PSIG.
_____	10.	Observe the compressor turn off and the low pressure begin to rise.
_____	11.	The compressor should turn on at 10 lb. Obtain a demonstration as required at this time in manually operating the low pressure cut-out.
_____	12.	Fine tune the range screw to obtain a cut-in or high event of 10 lb, calibrated to your gauge.
_____	13.	With the high event set at 10 lb, adjust the differential screw to obtain a cut-out of 0 lb, calibrated to your gauge.
_____	14.	Repeat as required to obtain correct cut-in and cut-out.
_____	15.	System operates correctly on low pressure safety function.

CALIBRATE SYSTEM FOR PRESSURE CYCLE DEFROST

Check	Step	Procedure
_____	1.	Raise cut-in until the pointer is at 40 lb by turning range screw.
_____	2.	Lower cut-out to 10 lb.
_____	3.	Front seat the receiver valve to lower evaporator pressure.
_____	4.	The compressor will turn off at approximately 10 lb.
_____	5.	Observe the low pressure gauge rise with the compressor off.
_____	6.	Crack the receiver service valve to add pressure.
_____	7.	Obtain the correct cut-in and cut-out for a typical walk-in from any equipment manufacturer's temperature/pressure chart.
_____	8.	Calibrate the high event with the range screw.

244

Check	Step	Procedure

_____ 9. Observe several compressor startups to verify correct cut-in. Allow pressure to climb slowly for accuracy.

_____ 10. High event cut-in occurs at _____ lb.

_____ 11. Install the lock plate to fix the cut-in.

_____ 12. Allow compressor to operate and observe low pressure gauge.

_____ 13. Adjust differential screw to obtain the desired cut-out.

_____ 14. Observe several compressor off cycles to verify accuracy.

_____ 15. Record the final cut-in and cut-out pressures.

Cut-in (high event) = _____   Cut-out (low event) = _____

_____ 16. Install the field adjustment knob on the differential adjustment screw in a midpoint position.

_____ 17. Instruct the customer in the use of the field adjustment knob which raises or lowers the cut-out by changing the differential and raises or lowers the average temperature in the space. Several cycles will be required to determine average storage temperature after an adjustment is made.

_____ 18. Explain that this means while they can control the temperature the control device is not a thermostat and controls the temperature only indirectly. It will take 3–4 hr after an adjustment to obtain the normal average temperature.

LOW SIDE REPAIR

STUDY MATERIAL
Chapter 18

LABORATORY NOTES

Most commercial duty equipment has a separate condensing unit consisting of a compressor, condenser, common base, and service valves. All fans and electrical components will also be stored on this common base. The liquid line leaving the condensing unit will have a king valve or a receiver service valve that can isolate this part of the system. A low side repair is the process of storing all the system refrigerant in the condensing unit and making a repair or component replacement to the rest of the system. This avoids the process of recovery and sometimes even evacuation for refrigerant side repairs.

The condensing unit is fabricated and tested at the factory as a separate component and seldom needs refrigerant side repairs. The trick in a low side repair is to balance the low side at 2–5 lb of positive pressure so that when the system is opened air will not be drawn into the system and only a minimal amount of refrigerant will be lost. This process is called pumping down the system. Since the liquid line goes to low pressure during the pump down process, it is considered a part of the low side during this repair. The liquid line is definitely a part of the high pressure side during normal operation.

UNIT DATA

1. System description, name or ID #_____

2. Repair to be made _____

PUMP DOWN PROCEDURE

Check	Step	Procedure
_____	1.	Obtain all tools, materials, and replacement parts.
_____	2.	Install gauges on suction and discharge service valves.
_____	3.	Observe the system operating in a normal operation.
_____	4.	Record operating pressures. High side pressure = _____ Low side pressure = _____
_____	5.	Locate and identify the receiver outlet service valve or the liquid line service valve to be used in the pump down procedure.

Check	Step	Procedure
_____	6.	Demonstrate the normal operation of equipment and the liquid line valve to your instructor.
_____	7.	Front seat the liquid line service valve.
_____	8.	Observe the low pressure drop off. If the system uses a low pressure cut-out for compressor cycling, the low pressure cut-out will have to be bypassed temporarily with a thin bladed screwdriver inserted below the spring. Obtain a demonstration on this process.
_____	9.	Turn off the compressor at about 6 in Hg vacuum on the low pressure side.
_____	10.	Observe the low pressure rise to above 0 PSIG.
_____	11.	Alternately run the compressor and open the liquid line service valve as required to balance the compound gauge at a steady low pressure of 2 lb of positive pressure.
_____	12.	The system is now ready for a repair anywhere from the liquid line valve to the compressor low pressure service valve.
_____	13.	To avoid the need for an evacuation, make sure that system exposure to the atmosphere is no longer than 1 min total.
_____	14.	Longer repairs such as leak repairs will require an evacuation of the low side only. All refrigerant will stay in the high side during the repair process and released back into the system when the repairs are completed.

SYSTEM REPAIR

Check	Step	Procedure
_____	1.	Describe in detail the repair as it was made (i.e., replace filter drier, move components, repair leaks, etc.). _____
		_____
		_____
_____	2.	Complete repairs. _____

RESTORE NORMAL SYSTEM OPERATION

Check	Step	Procedure
_____	1.	Close up lines and inspect for leaks.
_____	2.	For repairs of less than 1 min exposure, proceed to step 8.
_____	3.	Leak test as required, install leak testing mixture into low side only.
_____	4.	Evacuate from low side only.
_____	5.	Use one hour, triple vac, or micron evacuation as desired.
_____	6.	Perform vacuum pressure drop test as desired.
_____	7.	Remove vacuum pump with system maintaining 30 in Hg vacuum.
_____	8.	Turn liquid line service valve to cracked off the backseat position.
_____	9.	Observe pressure build up in low pressure side.

Check	Step	Procedure
	10.	Turn system on.
	11.	Observe normal system operation.

FINAL SYSTEM TEMPERATURES AND PRESSURES

Check	Step	Procedure
	1.	Record system operating pressures.

High side pressure = _____   Low side pressure = _____

Check	Step	Procedure
	2.	Check that the suction line is cool to the touch.
	3.	Check that the liquid line is warm to the touch.
	4.	Check that normal cooling appears to be in progress.
	5.	Observe normal cooling for 5 to 15 min. Check that the normal cooling is still OK.
	6.	Record final system operating pressures and space temperature.

High side pressure = _____   Low side pressure = _____   Temperature = _____

HIGH PRESSURE DOME STARTUP

STUDY MATERIAL
Chapter 5

LABORATORY NOTES

Most rotary compressors are high pressure domes, which means the compressor discharges to the welded steel and the space surrounding the compressor and motor is high pressure refrigerant during operation. For the opposite example, all piston type compressors are low pressure domes. For high pressure domes, the suction line enters the dome, passes through, and enters the compressor suction port while the compressor discharges directly to the dome. This greatly affects the system operation, especially during startup after a long off cycle. During the off cycle refrigerant will migrate to and collect in the compressor oil.

This happens in all hermetic type compressor motor assemblies but in high pressure domes some particular problems are created. High pressure domes prevent the problems of flooded starts, oil foaming, and loss of oil during startup, but cause some problems of their own. Refrigerant that has migrated to the dome during an off cycle will be trapped in the dome and not be in circulation until the heat of the compressor drives the refrigerant into circulation. The system will appear to be undercharged during this time. This operating characteristic extends the system stabilization time required. Window AC units stabilize rather quickly while upright domestic units utilizing a warm wall condenser take longer.

UNIT DATA

1. Unit make _____   Model # _____

2. Unit description _____

3. Application (circle one):   AC   High temperature   Medium temperature   Refrigerator/freezer

SYSTEM INSPECTION

1. Type of condenser (circle one):   Fan coil   Static   In-the-wall

2. Type of evaporator (circle one):   Fan coil   Cold shelf   Cold wall   Bare pipe

3. Type of metering device (circle one):   Capillary tube   TX   Other _____

4. List the three service valve locations: Discharge _____   Suction _____   Process tube _____

5. List exact weight of refrigerant charge in ounces from nameplate.   _____oz of R-_____

6. Predict normal operating pressures using the ambient temperature of +35°F and the coil temperature method.

   High side pressure = _____   Low side pressure = _____

SYSTEM OPERATION

Check	Step	Procedure
_____	1.	Install gauges and record system idle pressures.

High side pressure = _____   Low side pressure = _____

Check	Step	Procedure
_____	2.	Start up system and record system operating pressures at 2, 5, or 10 min intervals. Consult your instructor on interval for pressure recording, depending on system type.

	Time	High side pressure	Low side pressure
1.	_____	_____	_____
2.	_____	_____	_____
3.	_____	_____	_____
4.	_____	_____	_____
5.	_____	_____	_____
6.	_____	_____	_____
7.	_____	_____	_____
8.	_____	_____	_____

What happens to the operating pressures as the system runs longer? _____

_____

CHARGE BY SUPERHEAT—REFRIGERATOR/FREEZER

STUDY MATERIAL
Chapter 5

LABORATORY NOTES

Every capillary tube system has a critical charge, that is, an exact amount of refrigerant required for best operation. When a system is designed and tested at the factory, the charge amount is determined by test. The test used is to operate the system in a normal ambient temperature and to measure operating pressures, superheat, subcooling, and perhaps also Btu output, etc. The charge amount is recorded for the best performance of the system. This charge amount will be affected by any of the following conditions: field altered components, constant operation in abnormal ambient conditions, dirty coils, restricted airflow, etc.

We can duplicate this test in the field using an accurate remote lead thermometer and gauge manifold set. This is the most accurate way to check the charge on an altered system or any capillary tube system that you don't know the correct charge. The key is to duplicate the design or normal operating pressures that will be present in the system normal operating mode and to charge the system to the correct operating superheat. Done properly this charging method will duplicate the factory test and compensate the charge for any field changes that have been made. To verify what the correct superheat should be in a given system, evacuate and recharge the system using the by weight method of the original nameplate refrigerant. This refrigerant could be recovered and an alternate refrigerant installed, duplication the evaporator coil temperature, condensing temperature, and suction line superheat. I have found that correct operation can be obtained on domestic refrigerator/freezers with –20°F coil, 105°F condensing temperature, and 80°F superheat (+/–10°F on superheat). The superheat figure seems high at first, but remember the capillary tube is soldered to the suction line adding to the suction line temperature.

UNIT DATA

1. Unit make _____ Model # _____ Color _____

2. Unit type (circle one):   Freezer   Refrigerator/freezer

3. Box type (circle one):   Top freezer   Bottom freezer   Chest   Side-by-side

   Other _____

4. System refrigerant   Type _____ Amount _____ oz of R-_____

UNIT DESCRIPTION

1. Unit type (circle one):   Air spillage   Auto defrost   Refrigerant spillage   Cold wall

   In wall condenser

2. Defrost type (circle one):   Electric   Hot gas   Manual   Air cycle

RECOVERY AND EVACUATION

Check	Step	Procedure
_____	1.	Recover existing refrigerant charge. This being a Type 1 system, recovery can be to a bag.
_____	2.	Pressurize system to 25 PSIG R-22 and boost to 100 PSIG with nitrogen, helium, or CO_2 for a legal leak testing gas. Obtain a demonstration on the use of the nitrogen cylinder as required.
_____	3.	Evacuate system using 1 hr, triple vac, or micron method.
_____	4.	System should be left with gauges installed and system holding 30 in Hg vacuum.
_____	5.	Figure and record target temperatures and pressures using –20°F for the coil temperature, 105°F for the condensing temperature, and 80°F for the superheat.

Suction line temperature = 60°F – (–20) = 80°F superheat

High side pressure = _____   Low side pressure = _____

SYSTEM RECHARGE

Check	Step	Procedure
_____	1.	Obtain appropriate refrigerant cylinder and scale.
_____	2.	Weigh and record cylinder weight. _____lb _____oz
_____	3.	Hook center hose to refrigerant cylinder. Use vapor for R-12, R-22 etc. Charge all used/recycled and R-400 series refrigerant in liquid state.
_____	4.	Open refrigerant cylinder valve.
_____	5.	Purge center hose at manifold as required.
_____	6.	Open high side gauge handle putting refrigerant into high side of system.
_____	7.	Observe refrigerant pressure buildup on low pressure side of system.
_____	8.	Install about two thirds of listed by weight charge.
_____	9.	Turn system on. Observe high side pressure go up and low side pressure go down.
_____	10.	Let system run for 5 to 10 min.
_____	11.	Add refrigerant 1 oz at a time, 5 min between charges until cooling seems normal, consult your instructor at this point, because you don't want to get your system overcharged. Do not exceed system listed amount.
_____	12.	Once normal cooling has begun, let system run and cool for 1 to 4 hr, or until freezer reaches 10°F.
_____	13.	Install one thermometer lead in freezer, and one lead on suction line 1 ft from compressor.

Check	Step	Procedure
_____	14.	With freezer at 10°F and system still operating, superheat should be about 80°F (–20 coil and 60°F suction line).
_____	15.	Measure actual system operating superheat at this point by the following formula:

Suction line temperature – Coil temperature = Superheat

_____ – _____ = _____

Check	Step	Procedure
_____	16.	Allow system to operate for 24 hr.

FINAL SYSTEM READINGS

Check	Step	Procedure
_____	1.	Measure and record: High side pressure = _____ Low side pressure = _____
_____	2.	Measure and record temperatures. Fresh Food _____ Freezer _____
_____	3.	Measure suction line temperature and calculate final superheat. Superheat = _____
_____	4.	Record cylinder contents: _____ lb _____ oz
_____	5.	Record total refrigerant installed. Charge = _____ oz of R-_____

CHARGING BY SUPERHEAT, COMMERCIAL REACH-IN

STUDY MATERIAL
Chapter 18

LABORATORY NOTES

A commercial reach-in is a capillary tube system that, like every other capillary tube system, requires a critical charge. The term critical charge means it takes an exact amount of refrigerant to function at its peak performance. On newer units the charge weight is listed on the unit nameplate. If the system has been altered this listed charge will be wrong and the system will have to be charged by the superheat method. Even when the exact weight is listed and correct, it's a good idea to check the charge by the superheat method after sufficient run time for stabilization.

A commercial reach-in is a true refrigerator. There is no freezer in the system at all and the entire refrigerated space is a typical medium application requiring a medium temperature compressor and operating at approximately 10°F in the evaporator when cooled down. The superheat should run about 40°F with a 50°F suction line temperature and a 10°F coil. This will be true of all medium temperature reach-in boxes regardless of size or refrigerant type.

UNIT DATA

1. Unit make _____ Model # _____

2. System refrigerant. Type _____ Amount _____

3. Standard product and designed product temperature _____

SYSTEM INSPECTION OR PRESTART CHECKS

Check	Step	Procedure
_____	1.	Check/spin all fans for free rotation.
_____	2.	Oil all motors as required.
_____	3.	Inspect and clean condenser as required.
_____	4.	Predict normal operating pressures. High side pressure = _____ Low side pressure = _____

SYSTEM RECOVERY AND LEAK TEST

Check	Step	Procedure
_____	1.	Locate or install access valves on high and low side of system.
_____	2.	Install gauges and record system idle pressures.
		High side pressure = _____ Low side pressure = _____
_____	3.	Recover existing charge. Obtain refrigerant recovery Laboratory Worksheet R-2 as required.
_____	4.	Recover existing refrigerant pressure required by Federal guidelines, (10 in Hg). Record recovery pressure _____
_____	5.	Pressurize system to 25 PSIG R-22 and boost to 100 PSIG with nitrogen, helium, or CO2 for a legal leak testing gas. Obtain a demonstration on the use of the nitrogen cylinder as required.
_____	6.	Leak test with halide, soap bubbles, and electronic testers.
_____	7.	Record location of any leaks found.
_____	8.	System certified leak free and ready for evacuation.

EVACUATION

Check	Step	Procedure
_____	1.	Bleed leak testing mixture to atmosphere.
_____	2.	System is at 0 PSIG and ready for evacuation.
_____	3.	Use the 1 hr method, that is, 1 hr of evacuation time after 30 in Hg is reached on gauge.
_____	4.	Obtain electronic scale and refrigerant cylinder during evacuation time.
_____	5.	Perform vacuum pressure drop test as time is available.
_____	6.	Turn both gauge handles off.
_____	7.	Turn off and remove vacuum pump.
_____	8.	Connect charging hose to refrigerant cylinder.
_____	9.	Call for instructor to check system ready for charging.

RECHARGE

Check	Step	Procedure
_____	1.	System should be valved off and holding 30 in Hg vacuum.
_____	2.	Record weight of cylinder from electronic scale. _____ lb _____ oz
_____	3.	Open appropriate refrigerant cylinder valve. Use the liquid valve for any R-400 series or any recovered refrigerant, charge R-22, R-502, etc. in the vapor state.
_____	4.	Purge center hose at manifold.
_____	5.	Open high side pressure gauge manifold handle.

Check	Step	Procedure
_____	6.	Observe high side pressure gauge go up right away.
_____	7.	Observe low side pressure gauge go up slowly. You now know that refrigerant is traveling through the capillary tube.
_____	8.	Open the low side pressure gauge handle.
_____	9.	Install about 3/4 of the full charge into system.

SYSTEM STARTUP AND TEMPERATURE CHECK

Check	Step	Procedure
_____	1.	Turn system on.
_____	2.	Observe high side pressure go up and low side pressure go down.
_____	3.	Add refrigerant into low side to build system pressures until close to estimated operating pressures from above, or close to listed system weight of refrigerant is reached.
_____	4.	Allow system 15 min of run time.
_____	5.	Install thermometer lead in refrigerated space and on suction line 12 in upstream from compressor.
_____	6.	Operate system and add vapor slowly until the suction line temperature and space temperature begins to drop.
_____	7.	When space temperature reaches about 50°F, measure and record temperatures and pressures.

Space temperature = _____   Suction line temperature = _____

Low side pressure = _____

Check	Step	Procedure
_____	8.	Calculate superheat by using the following formula:

Suction line temperature – Low side coil temperature = Superheat

_____ – _____ = _____

FINAL TEMPERATURE PRESSURE CHECK
Record the following:

High side pressure = _____

Low side pressure = _____

Suction line temperature = _____

Comment on system operation. _____

_____

TYPICAL DOMESTIC REPAIR

STUDY MATERIAL
Chapter 5

LABORATORY NOTES

The typical domestic system (refrigerator, freezer, window AC unit, dehumidifier, etc.) will not have service valves installed at the factory. If any refrigerant side problems develop or are suspected in the field, these valves must be added. This is a major service job and should not be attempted before eliminating such common items such as dirty condenser, low airflow problems, defrost problems, etc.

Refrigerant side problems causing the complaint that the system runs but doesn't cool or that there is poor cooling could be: a partial loss of charge or inefficient compressor; or else a partial restriction in the filter, the capillary tube inlet screen, or the capillary tube itself. High and low side access valves are required to distinguish between these problems and make any repairs. The following repair procedure of refrigerant recovery includes installing access valves, replacing the filter drier, and recharging the system. These steps are a logical first step in servicing any system not equipped with factory service valves.

UNIT DATA

1. Shop ID # _____

2. Unit make _____   Model # _____

3. Unit description _____

4. System refrigerant.   Type _____   Amount _____

SYSTEM CHECK OUT

Check	Step	Procedure
_____	1.	Plug in system, listen for compressor startup.
_____	2.	Measure compressor amp draw. Verify that compressor is running within rated amperage.
_____	3.	Check at capillary tube for any cooling at all.
_____	4.	Check condenser inlet and first return bends for any signs of warm refrigerant gases circulating.
_____	5.	Inspect system for any existing service valves.

Check	Step	Procedure
_____	6.	Inspect system for a good location to install tee sections and Schrader valves as required.
_____	7.	Call for instructor check to verify poor cooling and need for refrigerant side repair and to check proposed location of access valves.

REFRIGERANT RECOVERY

Check	Step	Procedure
_____	1.	Select satisfactory service valve locations.
		High side valve location _____ Low side valve location _____
_____	2.	Install temporary tap-on type service valves.
_____	3.	Record system idle pressures. High side pressure = _____ Low side pressure = _____
_____	4.	Recover existing charge. Refer to Laboratory Worksheet R-2, Refrigerant Recovery, as required.

REPAIR PROCEDURE

Check	Step	Procedure
_____	1.	Sand and clean tee locations.
_____	2.	Obtain tees, Schrader valves, $1/4$ in tubing, torch, solder type filter and solder alloy. Use 15% sil-foss on copper to copper and silver solder on copper to steel or copper to brass.
_____	3.	Sand the inside of tees and Schrader stubs.
_____	4.	Cut lines slowly in the middle of the cleaned line.
_____	5.	Cut and remove old filter drier. Include a nominal 1 in of capillary tube removed with filter.
_____	6.	Notch and break capillary tube using file and needle nose pliers.
_____	7.	Inspect capillary tube end for debris or partial plug.
_____	8.	Remove cores from Schrader valves, store in flare caps.
_____	9.	Install and support, tees, $1/4$ in lines, valves, and filter.
_____	10.	Protect the floor, wires, etc., with wet cloths as required.
_____	11.	Light torch and braze all joints using sil-foss for all copper to copper joints and silver solder on copper to steel.
_____	12.	Inspect all brazed joints for good alloy flow.
_____	13.	Allow joints to cool and install gauge manifold.
_____	14.	Install refrigerant into high side of system.
_____	15.	Observe refrigerant pressure on compound gauge indicating a clear line through cap tube.
_____	16.	Install leak testing mix and leak test normally.
_____	17.	Perform system evacuation. [Optional] Remove cores if desired.
_____	18.	Obtain refrigerant and accurate scale.

Check	Step	Procedure
_____	19.	Put away soldering supplies, etc., during evacuation.
_____	20.	Predict normal operating pressures for system temperature range, refrigerant type, and ambient temperature operation. High side pressure = _____ Low side pressure = _____
_____	21.	Install holding charge. [Optional] Reinstall cores.
_____	22.	Install required refrigerant amount.
_____	23.	Start system and observe refrigeration begin.
_____	24.	Call for instructor inspection.

FINAL SYSTEM OPERATION CHECK

Check	Step	Procedure
_____	1.	Assuming normal operation, allow system to run and obtain normal space temperatures and cycle off.
_____	2.	Check during operation: High side pressure = _____ Low side pressure = _____
_____	3.	Record the following temperatures: Suction line temperature = _____ Suction superheat = _____ Ambient temperature = _____ Product space temperature = _____

CHECK AND ADJUST A TXV

STUDY MATERIAL
Chapter 5, Unit 5

LABORATORY NOTES

The function of a thermostatic expansion valve is to maintain a fully refrigerated evaporator over a wide range of operating pressures and conditions. It does this by sensing the suction line temperature and feeding more refrigerant as the suction line temperature rises. When the liquid refrigerant floods through the evaporator it cools the sensing element and closes the valve. The TXV then monitors evaporator coil superheat. Normally speaking a medium temperature system should be set to maintain 10°F of coil superheat. On low temperature equipment the superheat may be set lower, at about 5°F.

An externally equalized valve is a valve that is to be used on a system with over 2 lb of pressure drop across the evaporator. The manufacturer will determine this and equip the evaporator outlet with a fitting for an externally equalized valve. The fitting is a ¼ in SAE connection on the outlet of the coil that is connected to a ¼ in SAE connection on the side of the valve. The valve operates the same as an internally equalized valve but uses evaporator pressure from this ¼ in fitting rather than the pressure at the valve outlet.

Due to the time for system stabilization, in the field it is sometimes better for the customer to replace rather than to adjust a TXV, especially in lower temperature systems. In a school situation we have the time for you to experience this job, but we do it to learn how to identify a malfunctioning valve as much as to make repairs on installed valves.

UNIT DATA

1. Valve model # _____ System ID# _____

2. TXV make _____ Inlet size _____ Type _____

   Outlet _____

3. Temperature range _____

4. Equalized valve type (circle one): Internal External

MEASURE SYSTEM SUPERHEAT METHOD

Check	Step	Procedure
_____	1.	Install gauges on system.
_____	2.	Use an accurate remote lead thermometer. Install thermometer lead at the TXV remote bulb lead. Install the bulb on the suction line.

_____ 3. Operate system for 5 to 10 min, observing low side pressure and suction line temperature fluctuate, when they both seem to stabilize measure the suction line temperature.

Suction line temperature = _____

_____ 4. Measure low side pressure and read coil temperature from temperature/pressure chart for system

refrigerant type.   Coil temperature = _____

_____ 5. Calculate the operating suction superheat. Subtract your answer for #4 from #3.

Superheat = _____

_____ 6. Turn the adjustment stem in (clockwise) one 360 turn.

_____ 7. Observe low side pressure go down, not a lot but some.

_____ 8. Allow sufficient time for system stabilization.

_____ 9. Calculate the system operating superheat.   Superheat = _____

_____ 10. Turn the stem two turns counterclockwise.

_____ 11. Observe low side pressure go up.

_____ 12. Allow time for stabilization.

_____ 13. Calculate the system operating superheat.   Superheat = _____

_____ 14. Repeat as required to observe increase and decrease in refrigerant feed rate.

_____ 15. Adjust superheat to between 9°F and 11°F and install TXV stem cap.

_____ 16. Remove gauges and install all service caps.

FABRICATE A BASIC REFRIGERATION SYSTEM

STUDY MATERIAL
Chapter 5

LABORATORY NOTES

A basic refrigeration system includes the six parts of a system installed with the evaporator within an enclosed space. This procedure would be followed whenever installing the components of a walk-in or reach-in type refrigeration system. Most smaller systems are completely factory assembled while some units are sent remote, that is, with separate components to be assembled at the job site. A complete leak test, evacuate, and charge must be performed. In addition, the system must be wired according to electrical code requirements and must operate correctly.

SYSTEM ORGANIZATION

Check	Step	Procedure
_____	1.	Obtain the following components:
		a. Condensing unit
		b. Evaporator
		c. Cabinet
		d. TXV or capillary tube
_____	2.	Inspect job for refrigerant line routing and electric supply.
_____	3.	Position condensing unit for good airflow and ease of service electrically and mechanically.
_____	4.	Hang or install evaporator coil within cooled space or ducted to space.

SYSTEM PIPING

Check	Step	Procedure
_____	1.	Take measurements and route refrigerant line connections.
_____	2.	Obtain all required pipe, fittings, hangers, and insulation.
_____	3.	Install hangers for pipe support.

Check	Step	Procedure
_____	4.	Measure, cut, ream, and temporarily assemble pipe sections.
_____	5.	Sand or clean all pipe ends and fittings to be soldered.
_____	6.	Assemble, support, and braze; in that order.
_____	7.	Use sil-foss on all copper to copper joints and silver solder on any copper to steel or copper to brass.

LEAK-TEST, EVACUATE, AND RECHARGE

Check	Step	Procedure
_____	1.	Refer to standard leak test, evacuate, and recharge procedure.
_____	2.	Leak test with positive pressure first.
_____	3.	Bleed system to 0 PSIG before making any brazed leak repair.
_____	4.	Retest with pressure until leaks are eliminated.
_____	5.	System is leak free and ready for evacuation.
_____	6.	Bleed to 0 PSIG before connecting vacuum pump.
_____	7.	Evacuate system using either the 1 hr, triple vac, or micron method. Perform system wiring and complete paperwork during evacuation.
_____	8.	System must pass a 48 hr vacuum pressure drop test or maintain under 500 microns for 10 min before charging.
_____	9.	System has passed vacuum pressure drop test. 48 Hour test _____ Micron test _____
_____	10.	Call your instructor over for inspection before charging.
_____	11.	Break the vacuum with pure clean refrigerant.
_____	12.	Allow system to take cylinder pressure of vapor only.
_____	13.	Start system up and observe operation.
_____	14.	Use thin bladed screwdriver to bypass low pressure cut-out as required.
_____	15.	Use the sight glass method to determine correct charge.
		a. Normal pressures
		b Cool suction and warm liquid line
		c. Sight glass clears (bubbles will be seen during charging)
_____	16.	Record and label system for refrigerant type, weight installed, date, your name, and class.
_____	17.	Call instructor for system inspection.

TESTING SPLIT PHASE REFRIGERATOR COMPRESSORS

STUDY MATERIAL
Chapter 7, Unit 4

LABORATORY NOTES
This laboratory worksheet will follow the procedure for testing split phase refrigerator compressors.

OHMS TEST

Check	Step	Procedure
_____	1.	Obtain ohmmeter, start cord, ammeter, and a start capacitor of 88–108 microfarad x 120 V.
_____	2.	Determine compressor to be tested. It must be of the split phase or capacitor start type and 115 V to follow this procedure.
_____	3.	Expose the compressor terminals.
_____	4.	Using the ohmmeter R × 10 K scale, test compressor for ground by measuring ohms from each terminal to any section of bare copper line connected to the compressor.
_____	5.	Any measurable resistance from the windings to any line means the compressor is grounded and the test is completed.
_____	6.	If the compressor does not show a ground, proceed with the test and measure terminal to terminal on the R × 1 scale.
_____	7.	Zero the meter, draw a sketch of the terminal location, number the terminals, and record the exact resistance in the space provided.

C = Terminal left over while measuring the highest resistance

S = Common to the higher resistance

R = Common to the lowest resistance

Check	Step	Procedure

Ohm test 1 = 1 to 2 = _____ ohms   C = _____

Ohm test 2 = 2 to 3 = _____ ohms   R = _____

Ohm test 3 = 1 to 3 = _____ ohms   S = _____

_____ 8.   Repeat the ohmmeter test as often as required to obtain correct readings.

APPLIED VOLTAGE TEST

Record compressor nameplate rated load amps.   RLA = _____

Check	Step	Procedure
_____	1.	Use the start cord on a wooden bench only.
_____	2.	Don't plug in the cord until all other connections have been made and checked.
_____	3.	Measure amperage on the common leg at all times while the motor is energized.
_____	4.	Connect the start cord leads to common, start, and run according to the color code on the cord.
_____	5.	Connect the two capacitor leads to the start capacitor.
_____	6.	Be sure the rotary switch is in the off position.
_____	7.	Unlock the ammeter needle lock button.
_____	8.	Rotate the rotary scale to the 50 A scale.
_____	9.	Hook the ammeter around the common lead.
_____	10.	Plug in the start cord to 115 V.
_____	11.	Turn the rotary switch to the start position.
_____	12.	Drop back to the run position when compressor operation is verified.
_____	13.	Rotate the ammeter to a lower scale. Record amps. Amps = _____
_____	14.	Does compressor run within rated load amps (RLA) recorded above? (circle one): Yes No
_____	15.	If your answer to #14 was No, then turn off compressor.
_____	16.	Repeat the starting procedure as required to verify the compressor operation. 2nd _____ 3rd _____
_____	17.	Obtain a second compressor and repeat the above procedure.
_____	18.	Demonstrate the starting procedure for your instructor.

STARTING PERMANENT SPLIT CAPACITOR AND CAPACITOR START/CAPACITOR RUN COMPRESSOR MOTORS

STUDY MATERIAL
Chapter 7, Unit 4

LABORATORY NOTES

Residential AC units, whether split, package, or self-contained units, use permanent split capacitor (PSC) motors because of the high motor efficiency. A capacitor start/capacitor run (CSCR) compressor motor functions as a permanent split capacitor during run but has a start capacitor for greater start power. Both the PSC and the CSCR motors use a run capacitor in series with the start winding during run time to obtain their running efficiency. A typical start cord does not have a connection for the run capacitor. This type of start cord removes the start winding and start capacitor from the circuit, which would damage a PSC or a CSCR motor. If your start cord does not have a provision for a run capacitor, you must hand wire a run capacitor from R to S on the compressor, and put the start capacitor on the capacitor connection, which will be removed after the motor starts.

OHMS BENCH TEST

Check	Step	Procedure
_____	1.	Obtain ohmmeter, 230 V start cord, an assortment of start and run capacitors, assorted jumper wires, and an ammeter.
_____	2.	Select compressor #1 on test bench. Check with instructor if you are not sure if compressor is a PSC type.
_____	3.	Using the R × 10K ohm scale, measure and record ohms, terminal to frame. 1 = _____ 2 = _____ 3 = _____
_____	4.	Determine start, run, and common compressor terminals.

Check	Step	Procedure
_____	5.	Record ohm test results and sketch terminal location in space below.

C = Terminal remaining while measuring highest ohms

S = Terminal of higher resistance from common terminal

R = Terminal of lowest resistance from common terminal

Ohm test 1 to 2 _____   C = _____

Ohm test 2 to 3 _____   R = _____

Ohm test 1 to 3 _____   S = _____

APPLIED VOLTAGE TEST, PERMANENT SPLIT CAPACITOR

Check	Step	Procedure
_____	1.	Connect voltage supply wires to C and R. Do not plug in.
_____	2.	Hand wire an appropriate run capacitor from S to R terminals. Use manufacturer's recommended capacitor, or 15 or 20 microfarad for 1 HP, 25 microfarad for 2 HP, and 35 microfarad for a 3 HP. These run capacitors should be close enough for a short term bench test.
_____	3.	Hook ammeter on common lead.
_____	4.	Energize motor by plugging in or turning on power.
_____	5.	Observe motor operate and amps approach locked rotor amps (LRA) and drop to rated load amps (RLA).
_____	6.	If compressor fails to run or pulls high amps, check your wiring and call your instructor for an inspection.
_____	7.	Repeat starting process as required to read amperage.
_____	8.	Observe compressor pump air through ports.
_____	9.	Record amps during start in common wire. Amps = _____
_____	10.	Record amps during run in common wire. Amps = _____
_____	11.	Record amps during run in run wire. Amps = _____
_____	12.	Record amps during run in start wire. Amps = _____
_____	13.	Amperage in run wire and common wire should be the same.
_____	14.	The start winding amperage, measured in the wire going to the start terminal of the compressor, will be about 20–30% of the common amperage. Note that start winding amps is not to be confused with start amps or amps in the common lead during start.

APPLIED VOLTAGE TEST, CAPACITOR START/CAPACITOR RUN

Check	Step	Procedure
_____	1.	Addition of an appropriate start capacitor will convert any PSC motor to a CSCR.
_____	2.	Obtain 23 V start cord and appropriate start capacitor.
_____	3.	Connect 230 V start cord to C, S, and R as indicated on the start cord.
_____	4.	Hand wire in run capacitor from S to R terminals. Note that some 230 V start cords or boxes have both run capacitor and start capacitor connections at the box and do not require hand wiring in the run capacitor.
_____	5.	Install ammeter on common wire.
_____	6.	Plug in or connect to appropriate voltage supply.
_____	7.	Energize motor with start cord turning rotary switch to start and dropping back to run when the amps drop off and the compressor begins pumping air.
_____	8.	Record amp flow as follows:

Rated load amps (RLA) at common terminal: During start _____   During run _____

Run winding at R terminal: During start _____   During run _____

Start winding at capacitor: During start _____   During run _____

REPLACE A WELDED HERMETIC COMPRESSOR

STUDY MATERIAL
Chapter 18

LABORATORY NOTES

A hermetic compressor replacement is a very common job in the air conditioning and refrigeration service industry. Unlike the semihermetic replacement, a torch is needed due to the brazed connections. Because of the welded hermetic design, the exact cause of failure cannot always be pinpointed. A good rule is to perform a complete check of the system during startup to prevent repeat failures. Field replacement of a hermetic compressor should include a new filter drier in the liquid line. Any evidence of acid burnout should add a suction filter to the system. Standard refrigeration or air conditioning startup procedures should be followed to determine correct operation after the compressor replacement. It is not recommended to reuse refrigerant in a hermetic compressor replacement.

Check	Step	Procedure
_____	1.	Read and record compressor make and model #:
		Original _____ Replacement _____
_____	2.	Install gauges on system in the normal manner.
_____	3.	Record system idle pressures. High side pressure = _____ Low side pressure = _____
_____	4.	Attempt compressor operation to verify the failure and reason for replacement.
_____	5.	Demonstrate failure mode for customer as required.
_____	6.	Turn compressor off and lock out main disconnect.
_____	7.	Recover refrigerant to a separate cylinder and save cylinder for testing in case of acid indication.
_____	8.	Cut with a tubing cutter, suction, and discharge line in an accessible location. Twist off rota-lock fitting if used.
_____	9.	Remove compressor mounting bolts.
_____	10.	Draw a wiring diagram and identify all wires within the compressor terminal box.
_____	11.	Remove all wires entering terminal box.
_____	12.	Lift out compressor from condensing unit.
_____	13.	Using a scratch awl or prick punch, scribe a line on the top side of both suction and discharge line stubs.
_____	14.	Desolder suction and discharge line stubs.
_____	15.	Obtain an oil sample for inspection and acid testing.

Check	Step	Procedure
_____	16.	Braze suction and discharge line stubs in the same position as they were in old compressor. Keep scribed lines on top.
_____	17.	Lift and place in position replacement compressor.
_____	18.	Start all compressor bolts.
_____	19.	Using sil-foss and copper fittings solder suction and discharge line stubs to the system.
_____	20.	Install replacement filter.
_____	21.	Install leak testing mix of 25 lb R-22 inert gas boost.
_____	22.	Leak test compressor, filter, and all connections.
_____	23.	Evacuate system using recommended evacuation procedure.
_____	24.	Complete wiring connections, tighten compressor bolts, cleanup, and complete paperwork during evacuation.
_____	25.	Close gauge valves at manifold.
_____	26.	Add refrigerant to the high pressure side and observe refrigerant flow through to the low pressure side.
_____	27.	Charge by recommended charging procedure.
_____	28.	Startup system and obtain the following readings. Amps = _____ High side pressure = _____ Low side pressure = _____ Superheat = _____
_____	29.	Measure and record temperatures at the following: Suction coil outlet = _____ Suction at comp = _____ Liquid Line = _____
_____	30.	Calculate superheat and subcooling. Coil superheat = _____ Compressor superheat = _____ Liquid line subcooling = _____

REPLACE A SEMIHERMETIC COMPRESSOR
(MECHANICAL FAILURE ONLY)

STUDY MATERIAL
Chapter 18

LABORATORY NOTES

A semihermetic compressor using both suction and discharge bolt on service valves, can be replaced without the use of a torch. Typically a filter drier and sight glass is used, also with flare fittings. To replace a compressor that has not failed or a mechanical failure only, recovery or replacement of the refrigerant is not always necessary. We must remember that in the service world it would be unusual circumstances that would allow a compressor replacement without refrigerant recovery and replacement.

UNIT DATA

Check	Step	Procedure
_____	1.	Read and record compressor make and model #:
		Original _____ Replacement _____
_____	2.	Install gauges on system in the normal manner.
_____	3.	List temperature range and refrigerant type. Temperature range _____ R-_____

CHECK ORIGINAL COMPRESSOR

Check	Step	Procedure
_____	1.	Record system idle pressures. High side pressure = _____ Low side pressure = _____
_____	2.	Start up system and obtain the following readings: Amps = _____
		High side pressure = _____ Low side pressure = _____ Superheat = _____
_____	3.	Measure and record temperatures at the following locations.
		Suction line at coil outlet temperature = _____ Suction line at compressor temperature = _____
		Liquid line temperature = _____

Check	Step	Procedure
_____	4.	Calculate superheat and subcooling. Coil superheat = _____
		Compressor superheat = _____ Subcooling = _____
_____	5.	List any problems with the original compressor or system. _____

_____

_____

COMPRESSOR REPLACEMENT

Check	Step	Procedure
_____	1.	Turn off, lockout, and tag power to compressor.
_____	2.	Front seat both high side service valve and low side service valve.
_____	3.	Recover refrigerant from compressor crankcase only.
_____	4.	Draw a wiring diagram and remove all wires from compressor.
_____	5.	Remove suction and discharge valve bolts and pry off valve bodies.
_____	6.	Lift out old compressor from condensing unit.
_____	7.	Inspect and scrape clean as required both service valve gasket plates, (if gaskets are good, leave them be).
_____	8.	Lift and place in position replacement compressor.
_____	9.	Position both service valves and insert bolts.
_____	10.	Connect and tighten all refrigerant connections.
_____	11.	Pressurize compressor only on both high pressure and low pressure sides.
_____	12.	Leak test compressor, and all connections.
_____	13.	Evacuate compressor and all refrigerant connections.
_____	14.	Complete wiring connections, install compressor mounting nuts, clean up, and complete paperwork during evacuation.

SYSTEM STARTUP AND CHECK OPERATION

Check	Step	Procedure
_____	1.	With vacuum pump running, close both gauge handles.
_____	2.	Turn service valves from front seat to cracked off backseat.
_____	3.	Observe and record system idle pressures:
		High side pressure = _____ Low side pressure = _____
_____	4.	Start up system and obtain the following readings: Amps = _____
		High side pressure = _____ Low side pressure = _____ Superheat = _____

Check	Step	Procedure

_____ 5. Measure and record temperatures at the following locations.

Suction line at coil outlet temperature = _____

Suction line at compressor temperature = _____   Liquid line temperature = _____

_____ 6. Calculate superheat and subcooling.

Coil superheat = _____   Compressor superheat = _____   Subcooling = _____

_____ 7. Manually pump system down and replace liquid line flare filter.

_____ 8. Verify system normal operation.

MICRON EVACUATION

STUDY MATERIAL
Chapter 6, Unit 3

LABORATORY NOTES

This lab can only be done on a system that is clean, dry, and free from leaks. A good quality vacuum pump with freshly changed oil is required. A deep vacuum gauge with a clean sensor and good battery is also required. Your gauges and manifold must be free from leaks. Hoses must be in good shape and tested to hold a micron quality evacuation. One bad hose can cause you to not pass the micron vacuum pressure drop test. Even hoses that are perfectly suitable for charging can cause a failure in the micron vacuum pressure drop test. It is possible for a tiny leak through the wall of hose to be the problem. It is also possible in some newer hoses to outgas from the rubber enough gas to build up some pressure inside the hose or system and cause a failure of the vacuum pressure drop test. As in all evacuation processes, sometimes the Schrader valves are removed for a faster evacuation. However, especially when using a micron vacuum gauge, removing the cores is not really necessary because we are monitoring the vacuum level in microns and will know the quality of vacuum. Removing the cores is optional. I have noted the time at which the cores should be removed and put back in if you or your instructor choose to do this. Consult your instructor before removing Schrader cores and be sure to put them back in if you take them out.

UNIT DATA

1. Shop ID # _____

2. Unit description _____

3. System application and refrigerant type _____

4. Factory test pressures.   High side pressure = _____   Low side pressure = _____

PREPARATION

Check	Step	Procedure
_____	1.	Recover existing refrigerant as required.
_____	2.	Final leak test as required.
_____	3.	System leak free and ready for micron evacuation.

MICRON EVACUATION

Check	Step	Procedure
_____	1.	Check vacuum pump.
_____	2.	Obtain good quality vacuum pump and micron gauge.
_____	3.	Change oil in vacuum pump as required.
_____	4.	Connect vacuum pump to micron gauge only.
_____	5.	Check operation of pump and gauge by pulling a vacuum on sensor.
_____	6.	Check that vacuum reaches 200 microns or lower.

CHECK GAUGE MANIFOLD

Check	Step	Procedure
_____	1.	Connect gauge center hose to vacuum pump.
_____	2.	Open both manifold gauge handles.
_____	3.	Connect high pressure and low pressure hose to spuds on manifold.
_____	4.	Turn vacuum pump on and pull down to 200 microns or lower.
_____	5.	Close blank off valve on pump and observe vacuum gauge.
_____	6.	Manifold holds 500 microns or lower.

MICRON EVACUATION

Check	Step	Procedure
_____	1.	Bleed leak test mixture from system to 0 PSIG.
_____	2.	[Optional] Remove Schrader cores (as required), or move service valve stems to the intermediate position.
_____	3.	Install micron gauge in vertical position preferable at vacuum pump and with a separate valve.
_____	4.	Begin evacuation and observe pressure dropping.
_____	5.	Allow 15 min of evacuation time before exposing vacuum gauge sensor. This will keep oil from depositing on the micron sensor.
_____	6.	Open valve to micron gauge sensor and observe pressure dropping.
_____	7.	Record lowest pressure reached. Pressure = _____ microns
_____	8.	Close both gauge handles to determine quality of evacuation microns pump is now pulling. Pressure = _____ microns
_____	9.	Open gauge handles and continue evacuation of system until system vacuum bottoms out. Pressure = _____ microns
_____	10.	When system reaches approximately 300 microns you are ready to perform a vacuum pressure drop test.

282

MICRON GAUGE, VACUUM PRESSURE DROP TEST

Check	Step	Procedure

_____ 1.   Record pressure while vacuum is in progress.   Pressure = _____ microns

_____ 2.   Close blank off valve leaving system exposed to micron pressure gauge.

_____ 3.   Turn off vacuum pump.

_____ 4.   Observe micron pressure gauge seek true level of microns in system.

_____ 5.   Record microns after 10 minutes.   Pressure = _____ microns

_____ 6.   Use the following criteria to determine system condition:

      _____ a. A clean dry system will hold 500 microns or less.

      _____ b. A system with some moisture or contamination mixed with the system oil will level out as somewhere between 1000 and 1500 microns.

      _____ c. A system with a leak will have first the micron level rise continuously and then the gauge manifold low pressure gauge begin to go up.

      _____ d. A system with free water (water not mixed with oil) will climb off the scale on the micron gauge while the compound gauge stays at 30 in Hg.

_____ 7.   System passes pressure drop test if it does not climb above 500 microns within 10 min.

_____ 8.   Systems not passing can be cleaned up by the following steps.

      _____ a. Repeat leak testing.

      _____ b. Replace filter drier.

      _____ c. Change compressor oil.

      _____ d. Combine the triple vac with the micron method.

      _____ e. Change oil in the vacuum pump.

      _____ f. Retesting the vacuum pump, gauge manifold, and hoses to verify that your equipment will hold 500 microns for the 10 min vacuum pressure drop test.

      _____ g. Replace the hoses or the manifold as required.

_____ 9.   System certified to have maintained under 500 microns for 10 minutes.

REPLACE A SEMIHERMETIC COMPRESSOR (MILD BURNOUT, NO ACID)

STUDY MATERIAL
Chapter 18

LABORATORY NOTES

A semihermetic compressor has bolt on service valves to aid in the ease of replacement. When a flared filter drier is used, a complete compressor changeout can be performed without lighting a torch, no soldering required. In the case of a typical motor burnout, the repair procedure is much the same as a welded hermetic compressor replacement. The refrigerant will be recovered and new refrigerant installed in the system with the new compressor. An oil sample of the old compressor should be taken and tested to ensure there is no acid to deal with. Normal new system cleanup procedures of thorough evacuation and installing an oversized liquid line filter will be included.

Compressor replacement is a major expense and not to be done unnecessarily. Every compressor failure should be verified. If you are the one who condemned the original, then you should have it double-checked it before condemning. Some companies have a policy of not replacing a compressor without two service technicians independently testing it. The following procedure will be the situation if you have the new compressor on the truck but you did not condemn the old one. You must verify the failure mode before replacing the old one.

VERIFY FAILURE OF ORIGINAL COMPRESSOR

Check	Step	Procedure
_____	1.	Describe problem with old compressor. _____
		_____
		_____
_____	2.	Install gauges on system.
_____	3.	Record system idle pressures. High side pressure = _____ Low side pressure = _____
_____	4.	Inspect all wires and start components.
_____	5.	Perform ohms test of compressor (circle one): Ground Short
_____	6.	Attempt compressor operation to verify and record failure mode, if possible.
_____	7.	Demonstrate failure mode for customer as required.

Check	Step	Procedure
_____	8.	Record make and model number of existing and new compressor.
		Original _____ Replacement _____

REMOVE OLD COMPRESSOR

Check	Step	Procedure
_____	1.	Turn compressor off and lock out main disconnect.
_____	2.	Recover system refrigerant.
_____	3.	Vent compressor crankcase to atmosphere from suction service port or any other crankcase port.
_____	4.	Remove the four compressor mounting bolt nuts.
_____	5.	Verify accuracy or draw a wiring diagram to identify all wires within the compressor terminal box.
_____	6.	Disconnect and remove all wires entering terminal box.
_____	7.	Loosen and remove service valve bolts and any pressure lines, oil lines, etc.
_____	8.	Lift out compressor from condensing unit.
_____	9.	Perform acid test on oil sample from failed compressor.
		(circle one): No acid Acid present
_____	10.	If acid shows up, more cleanup will be required.
_____	11.	Pull head and look for signs of valve damage and overheating.

INSTALL NEW COMPRESSOR

Check	Step	Procedure
_____	1.	Remove service valve blocks from new compressor.
_____	2.	Inspect and clean service valve gasket surface as required.
_____	3.	Check oil level in new compressor. Fill to center of oil inspection glass, as required.
_____	4.	Lift replacement compressor into position.
_____	5.	Connect and tighten service valves and all pressure lines.
_____	6.	Install new or replace liquid line filter drier and M/F sight glass with flare fittings.
_____	7.	Pressurize system first on the high pressure side and then on the low pressure side.
_____	8.	Leak test all connections and the entire system.
_____	9.	Move service valves to midseat position for evacuation. List evacuation method. _____
_____	10.	Complete wiring connections and install compressor bolt nuts during evacuation.

RECHARGE AND START UP NEW SYSTEM

Check	Step	Procedure
_____	1.	Observe system vacuum in progress or system holding vacuum pressure drop test at 30 in Hg.
_____	2.	Close both gauge handles.
_____	3.	Move service valve stems to cracked off backseat.
_____	4.	Remove charging hose from vacuum pump and connect to virgin refrigerant of correct type. List refrigerant type. R- _____
_____	5.	Obtain scale and record weight of refrigerant cylinder. _____ lb _____ oz
_____	6.	Open cylinder valve and purge charging hose at manifold.
_____	7.	Open high side manifold gauge and begin adding refrigerant.
_____	8.	Observe low side pressure go up (pump down cycle system will keep refrigerant in the high side with solenoid closed).
_____	9.	System is a pump down cycle system and low side pressure must be put into the low side.
_____	10.	Turn system on and observe high side pressure go up and low side pressure go down.
_____	11.	Measure and record compressor amps. Rated full load amps = _____ Measured amps = _____
_____	12.	Pin low pressure cut-out closed with thin bladed screwdriver to prevent short cycling of compressor as required.
_____	13.	Continue running and adding refrigerant into low side until sight glass clears or normal high side pressure is reached.
_____	14.	Allow 10–15 min of run time while normal cooling is reached.
_____	15.	Add refrigerant as required to clear sight glass at a full load condition.
_____	16.	Record weight of cylinder at full charge. _____ lb _____ oz
_____	17.	Calculate total refrigerant charged by using the following formula:

#5 (above) _____ – #16 (above) _____ = _____ lb _____ oz

FINAL SYSTEM TEMPERATURE AND PRESSURE

Check	Step	Procedure
_____	1.	Measure and record the following. High side pressure = _____ Low side pressure = _____ Compressor full load amps = _____
_____	2.	Measure suction line temperature at the following locations. One foot from compressor = _____ At coil outlet = _____

Check	Step	Procedure

 _____   3.   Calculate superheat at the following locations.

Compressor = _____   TXV or coil superheat = _____

 _____   4.   Liquid line temperature at the following locations.   Sight glass = _____   At TXV = _____

 _____   5.   What, if anything, was done to correct the original failure mode from recurring?_____

_____

_____

REPLACE A SEMIHERMETIC COMPRESSOR
(ACID BURNOUT)

STUDY MATERIAL
Chapter18

LABORATORY NOTES

An acid burnout is the most severe compressor burnout problem that can occur. Moisture must have been present to help create the acid. High operating temperature for an extended period of time combined with the moisture creates the acid. When a compressor is replaced without getting the acid out, a repeat failure can occur overnight. Some people claim they can smell acid in refrigerant and oil but most people need to perform an acid test. Oil samples when mixed with the appropriate chemicals will change color. Several brands of acid test kits are available. The amount of oil required and the colors will vary. An oil sample saved in a small clean bottle can be taken at the site and tested later. Signs of acid damage are any severe burnout, copper plating on the head and valve plate, and the compressor showing .5 Megohm or lower on an insulation resistance test. If acid is suspected, it is recommended to be tested and verified with an acid kit.

To get rid of acid requires more than just new oil in the new compressor, and an oversized filter drier. A permanent suction filter is sometimes installed in smaller systems, in addition to the oversized liquid line filter. Larger systems and severe acid contamination use a 48 hr filter change, oil change, and an additional acid test. This process will be repeated every 48 hr until the acid test show no acid. A cleanout kit, sometimes used on larger systems, is composed of a replaceable core suction filter and a three-way bypass valve assembly on the filter. It allows the service technician to replace the suction cores without shutting the system down. It also allows for a variety of cores to be used. High acid, standard, and plain felt cores are available.

VERIFY FAILURE OF ORIGINAL COMPRESSOR

Check	Step	Procedure
_____	1.	List or describe problem with old compressor. _____ _____ _____
_____	2.	Install gauges on system in the normal manner.
_____	3.	Record system idle pressures. High side pressure = _____ Low side pressure = _____
_____	4.	Inspect all wires and start components.

_____ 5. Measure and record ohms test and insulation resistance test ohms or megohms to ground.

Resistance = _____

_____ 6. Attempt compressor operation to verify and record failure mode.

_____ 7. Demonstrate failure mode for customer as required.

_____ 8. Record make and model number of existing and new compressor.

Original _____   Replacement _____

REMOVE OLD COMPRESSOR

Check	Step	Procedure

_____ 1. Turn compressor off and lock out main disconnect.

_____ 2. Recover system refrigerant.

_____ 3. Vent compressor crankcase to atmosphere from suction service port or any other crankcase port.

_____ 4. Remove the four compressor mounting bolt nuts.

_____ 5. Verify accuracy or draw a wiring diagram to identify all wires within the compressor terminal box.

_____ 6. Disconnect and remove all wires entering terminal box.

_____ 7. Loosen and remove service valve bolts, and any pressure lines, oil lines, etc.

_____ 8. Lift out compressor from condensing unit.

_____ 9. Perform acid test on oil sample from failed compressor.

(circle one):   No acid   Acid present

_____ 10. Pull head. Check for valve damage and/or overheating.

Since acid has been shown to be present for this procedure to be followed, we must deal with the acid. The following lists five generic levels of system cleanup generally followed by service technicians, in an increasing order of severity in contamination. Choose one of the following that applies to your situation:

_____ 1. Level 1 = Oversized liquid line filter.

_____ 2. Level 2 = Oversized liquid line filter and permanently installed suction line filter.

_____ 3. Level 3 = Oversized liquid line filter and replaceable core suction filter, replace cores as required.

_____ 4. Level 4 = Oversized liquid line filter and replaceable core suction filter and 48 hr oil and filter change.

_____ 5. Level 5 = Oversized liquid line filter and multiple core suction filter cleanout kit with 48 hr oil and filter change.

INSTALL NEW COMPRESSOR

Check	Step	Procedure
_____	1.	Exchange service valve blocks from new compressor.
_____	2.	Inspect and clean service valve gasket surface as required.
_____	3.	Check oil level in replacement compressor fill to center of oil inspection glass.
_____	4.	Lift and install replacement compressor.
_____	5.	Connect and tighten service valves and all pressure lines.
_____	6.	Install oversized liquid line filter.
_____	7.	Install suction filter or clean-out kit. (circle one): Suction filter Clean-out kit
_____	8.	Pressurize system first on the high pressure side and then on the low pressure side.
_____	9.	Leak test all compressor and filter drier connections.
_____	10.	Leak test entire system as required.
_____	11.	Evacuate system and record vacuum method.
		(circle one): One hour Triple vac Micron test
_____	12.	Complete wiring connections and install compressor bolt nuts during evacuation.

RECHARGE AND STARTUP NEW SYSTEM

Check	Step	Procedure
_____	1.	Observe system vacuum in progress or system holding vacuum pressure drop test at 30 in Hg.
_____	2.	Close both gauge handles.
_____	3.	Move service valve stems to cracked off backseat.
_____	4.	Remove charging hose form vacuum pump and connect to virgin refrigerant of correct type.
		Refrigerant type: R- _____
_____	5.	Obtain scale and record weight of refrigerant cylinder. _____ lb _____ oz
_____	6.	Open cylinder valve and purge charging hose at manifold.
_____	7.	Open high side manifold gauge and begin adding refrigerant.
_____	8.	Observe low side pressure go up. Note that pump down cycle system will keep refrigerant in the high side with solenoid closed.
_____	9.	System is a pump down cycle system and low side pressure must be put into the low side.
_____	10.	Turn system on and observe high side pressure go up and low side pressure go down.
_____	11.	Measure and record compressor amps.
		Rated full load amps = _____ Measured amps = _____
_____	12.	Pin low pressure cutout closed with thin bladed screwdriver to prevent short cycling of compressor as required. (circle one): Yes No

Check	Step	Procedure
_____	13.	Continue running and adding refrigerant into low side until sight glass clears or normal high side pressure is reached.
_____	14.	Allow 10–15 min of run time while normal cooling is reached.
_____	15.	Add refrigerant as required to clear sight glass at a full load condition.
_____	16.	Record weight of cylinder at full charge. _____ lb _____ oz
_____	17.	Calculate total refrigerant charged by using the following formula:

#5 (above) _____ – #16 (above) _____ = _____ lb _____ oz

FINAL SYSTEM TEMPERATURES AND PRESSURES

Check	Step	Procedure
_____	1.	Final system operating pressures. High side pressure = _____ Low side pressure = _____
_____	2.	Measure and record the following information.

    a. Compressor full load amps = _____

    b. Suction line temperature at 1 ft from compressor = _____   Temperature at coil outlet = _____

    c. Suction line superheat = _____   TXV or coil superheat = _____

    d. Liquid line temperature at sight glass = _____   At TXV = _____

FORTY EIGHT HOUR OIL AND FILTER CHANGE

Check	Step	Procedure
_____	1.	Allow two days of normal operation.
_____	2.	Install gauges on discharge line service valve and suction side service valve.
_____	3.	Pump system down manually by front seating liquid line service valve or receiver service valve.
_____	4.	Verify system at 2 lb to 5 lb positive pressure.
_____	5.	Manually turn off power to compressor.
_____	6.	Front seat discharge service valve.
_____	7.	Use drain plug, suction gun, or vacuum suction kit, to remove oil from compressor.
_____	8.	Perform acid test on oil removed. (circle one): No acid Acid present
_____	9.	Removed oil shows no acid. The cleanup process is complete.
_____	10.	Replace liquid line filter.

Check	Step	Procedure
_____	11.	Change cores in cleanout kit, use high acid cores if acid is still present. If there is no acid present use standard felt cores or leave empty. Alternately, remove temporary suction filter. Describe.
		_____
		_____
_____	12.	Evacuate entire low side of system.
_____	13.	Turn discharge line service valve to the cracked off backseat position.
_____	14.	Turn receiver service valve to backseat position.
_____	15.	Repeat as required to get rid of all traces of system acid.
_____	16.	System free from acid and running normal.

SEMIHERMETIC COMPRESSOR TEARDOWN

STUDY MATERIAL
Chapter 18

LABORATORY NOTES
This laboratory worksheet will follow the procedure of a semihermetic compressor teardown.

COMPRESSOR DATA

1. Unit make _____ Model # _____ Serial # _____

2. Voltage _____ Phase _____ Refrigerant type _____ Temperature range _____

3. Compressor cooling system type (circle one): Air Refrigerant

5. Ohms test.

   Terminal Sketch

   Single phase

   Ohms 1–2 = _____

   Ohms 2–3 = _____

   Ohms 1–3 = _____

   Ohms terminal to frame = _____

   Three phase

   Ohms 1–2 = _____

   Ohms 2–3 = _____

   Ohms 1–3 = _____

   Ohms terminal to frame = _____

MEGGER TEST
It should be pointed out that the instructions that come with most meggers are intended for open motors and do not apply to refrigeration compressor motors. Copeland's Application Engineering Bulletin # AE-1294 addresses this problem. Using the megger open motor criteria to condemn a refrigeration compressor would be a mistake.

Check	Step	Procedure
_____	1.	Obtain insulation resistance tester from cabinet.
_____	2.	Connect meter from frame to winding.
_____	3.	Record reading. Resistance = _____ megohms

Megohm values and diagnoses:

- 20+ megohms signifies that the motor in good condition.

- 20 to .5 megohms shows signs of moisture in oil. It is recommended to change filter and oil.

- .5 megohms or less signifies damage to the motor winding. The appropriate course is to condemn the compressor.

Check	Step	Procedure
_____	4.	Conclusion from ohmmeter test and insulation resistance test. Compressor is
		(circle one): Good Grounded Shorted
_____	5.	If the motor is shorted or grounded, the job is done. If the motor checks out good by the ohmmeter test, proceed to the applied voltage test. Comment. _____

APPLIED VOLTAGE TEST

This test is set up for either a split phase, capacitor start, PSC, CSCR, or three phase motor.

Check	Step	Procedure
_____	1.	Refer to unit data or nameplate for correct voltage.
_____	2.	Locate correct voltage source.
_____	3.	Move compressor to voltage source. Preferably the test should be performed on a wooden table.
_____	4.	Vent discharge and suction lines for free flow of air.
_____	5.	Connect wires at appropriate terminals for motor type. Use appropriate start cords or jumpers as required.
_____	6.	Install ammeter on common lead for single phase or any lead of a three phase motor.
_____	7.	Support the motor and observe ammeter during motor startup. The motor current will rise and approach locked rotor amps during startup but drop off quickly if the motor starts and runs.
_____	8.	Energize motor.
_____	9.	Repeat as required.
_____	10.	Conclusion. Motor diagnosis is as follows:

_____   a. Seized up, motor will not turn over at all.

_____   b. Mechanical failure, motor turns over with loud clanking.

_____   c. Tight bearings, starts hard, runs hard, draws high amps.

_____   d. Shorted turn to turn, blows fuses.

_____   e. Partly shorted, runs easy but with high amperage. Comment._____

It must be remembered that a compressor failure is usually the result of a system problem. The system problem can be identified by analyzing the compressor motor assembly. Keep in mind that the compressor portion is a separate component from the compressor motor, even though they are connected by a common crankshaft. A compressor motor failure in addition to compressor damage indicates a long term problem or failure during operation. A motor failure only with no damage to the compressor indicates a short term failure during compressor idle time, such as lightning, miswiring, electrical supply, or electrical component failure. A compressor failure with no damage to the motor winding or system contamination means a relatively quick failure such as something coming apart. A compressor teardown is performed to distinguish between the various possible causes of mechanical or electrical failure. System contamination resulting from burnout (if any) must be removed in any case. System problems such as TXV overfeeding, TXV underfeeding, off cycle migration, incorrect piping, etc., must be corrected or the replacement compressor will be in jeopardy.

MOTOR TEARDOWN

Motor teardown must be performed on defective compressors.

Check	Step	Procedure
_____	1.	Move compressor to convenient workbench.
_____	2.	Vent compressor crankcase, suction line, and discharge line.
_____	3.	Remove oil fill plug and drain oil into a clean container.
_____	4.	Record volume of oil removed.
_____	5.	Measure or estimate amount of oil in ounces. Volume = _____ oz
_____	6.	Inspect oil for color, odor, and contamination.
_____	7.	Perform acid test as desired.
_____	8.	Record results of acid test. Acid content = _____
_____	9.	Inspect discharge line for signs of contamination.
_____	10.	Break loose all head bolts.
_____	11.	Remove all but two head bolts.
_____	12.	Bag and identify all bolt groups and individually distinctive bolts for the reassembly process.
_____	13.	With two head bolts in place, break loose head.
_____	14.	Remove compressor head and valve plate.
_____	15.	Inspect head for the following indicators.

_____   a. Carbon and sludge. These indicate temperatures of up to 300°F.

_____   b. Coke varnish. This indicates compressor overheating above 325°F. Coke varnish looks like copper plating but can be scraped off while true copper cannot.

_____   c. Rainbow valve plate. This condition indicates overheating above 350°F.

_____   d. Copper plating. This indicates overheating in the presence of acid. If you suspect copper plating, test oil to verify the presence of acid.

| _____ | 16. | Inspect for valve damage or other signs of slugging. |

_____   a. Broken suction reeds are due to oil slugging.

_____   b. Broken discharge reeds are due to liquid refrigerant slugging.

Check	Step	Procedure
_____	17.	Push down on pistons. Free play in pistons indicates rod bearing wear. In this case, the compressor may have been knocking during operation.
_____	18.	Remove motor end plate and compressor end plate.
_____	19.	Inspect end bearings for signs of wear.
_____	20.	Roll compressor on its side and remove bottom plate.
_____	21.	Inspect rod bearings for wear by turning rotor manually back and forth.
_____	22.	Remove rod bearings, crankshaft, and pistons.

_____ a. Remove rotor end bolt.

_____ b. Remove rotor.

_____ c. Hold pistons in position while pushing or driving crankshaft out. Use a wood or brass rod when driving crank out. Do not hammer on crank with a steel hammer or shaft.

_____ d. Remove crankshaft.

_____ e. Drop out pistons.

Check	Step	Procedure
_____	23.	Inspect piston rings or top edge for knife edge condition.
_____	24.	Inspect piston walls for vertical streaking.

_____ a. Good pistons will still have a gray surface, which is lubrite. Lubrite is a lubricating coating from the factory.

_____ b Vertical streaks indicate excessive wear.

Check	Step	Procedure
_____	25.	Inspect rod bearing surfaces for wear or breakage.

_____ a. Rod bearing wear in all bearings will cause flooded starts.

_____ b. Rod bearing wear in one end only will cause continuous floodback.

_____ c. Severe rod bearing wear is caused by lack of oil.

Check	Step	Procedure
_____	26.	Inspect cylinder walls for discoloring or scoring.

Comments. _____

_____

REASSEMBLY

Check	Step	Procedure
_____	1.	Assemble all parts in reverse order.
_____	2.	Start all bolts in each plate before tightening.
_____	3.	Snug all bolts.
_____	4.	Clean area and wipe oil from compressor.
_____	5.	Consult your instructor for compressor identification and storage.

1. COMPRESSOR OVERHEATING/LONG TERM BURNOUT

A normal amount of compressor oil that is dark and strong smelling but without acid coupled with signs of bearing wear indicates a prolonged compressor overheating problem. The system should be inspected for a dirty condenser and other causes of high head pressure or high suction superheat. Overheating may be caused by an inefficient compressor valve with excessive rod bearing wear or broken valves. Severe cases of overheating in the presence of moisture will cause acid in the system, creating acid burnout, the most difficult burnout problem to get rid of.

2. LACK OF LUBRICATION

Loss of oil to the system, or lack of oil in the compressor for any reason, will cause excessive wear on moving parts and bearings. On suction cooled compressors the rear main will show wear and cause the rotor to drop in the increased tolerance. Eventually the rotor will drop low enough to come in contact with the stator and rub or cut the winding, causing a spot burn.

3. SLUGGING (REFRIGERANT OR OIL)

A compressor that slugs liquid refrigerant or oil will show valve damage. Broken valves without signs of other wear patterns identify slugging as the cause of the problem.

4. REFRIGERANT FLOODBACK

Regular liquid floodback during operation will show up as a damaged bearing. The first bearing the refrigerant comes in contact with will show the wear. Any severe wear in a single bearing or area, with no damage to other areas, is a sign of floodback. In a suction cooled semihermetic, the center bearing is the one subject to floodback. This bearing wear will cause a rotor drop and accompanying motor burnout. This is a prime example of an electrical failure that was really caused by a mechanical problem. If the floodback problem is not corrected, another electrical failure will be in order.

5. FLOODED START

Regular startups with refrigerant in the motor will cause loss of oil to the system and result in a temporary loss of lubrication. This will result in excessive overall bearing wear primarily in the center rod bearings without sign of oil slugging or long term overheating. Valve plate damage may also be present.

PERFORMANCE TEST A SEMIHERMETIC COMPRESSOR

STUDY MATERIAL
Chapter 18

LABORATORY NOTES

This lab involves a complete test of an operating compressor to verify its field performance. It is the most accurate field procedure for identifying inefficient compression or a partial short in the motor winding. An example of inefficient compression would be the compressor running but not quite pumping enough refrigerant. A partial short turn to turn within the motor winding would cause a higher than normal amperage with less torque delivered to the compressor. Both of these could be the cause of a poor cooling complaint from the customer.

The procedure requires accurate measurement of system pressures and amperage at those pressures. The readings can be taken and plotted on the compressor performance chart later. Or a copy of the chart can be taken to the site and done on the job. The chart is a plotted curve of the original factory test amperage at any set of operating conditions. This chart is available from most manufacturers. If the compressor is pulling a high amperage for the pressures, it is a partial short. If the compressor is pulling low amperage for the pressures it is inefficient or has bad valves.

UNIT DATA

1. Compressor make _____   Model # _____

2. Refrigerant type _____   Temperature range _____

SYSTEM INITIAL INSPECTION

Check	Step	Procedure
_____	1.	Check/spin all fans for free rotation.
_____	2.	Condenser coils are clean and have normal airflow.
_____	3.	Evaporator coils are clean and have normal airflow.
_____	4.	Check all electrical connections for loose wires.
_____	5.	Install gauges on high and low side of system.

OPERATE COMPRESSOR AND MEASURE DATA

Check	Step	Procedure
_____	1.	Obtain a normal system operation.
_____	2.	Allow 15 min of continuous operation for system stabilization.
_____	3.	Install ammeter on common lead measuring compressor amperage only.
_____	4.	Measure and record amperage and system pressures. Compressor amps = _____
		High side pressure = _____ Low side pressure = _____

PLOT DATA ON PERFORMANCE CURVE OF COMPRESSOR

Check	Step	Procedure
_____	1.	Obtain performance curve of compressor by model number.
_____	2.	Locate low side pressure at bottom of chart.
_____	3.	Draw a vertical line up from low side pressure crossing high side pressure line on the amps portion of the chart.
_____	4.	Read expected amperage from chart. Amps = _____
_____	5.	Is actual amperage higher than, less than, or the same as the expected amperage you read from the chart? (circle one): Higher than Lower than Same as
_____	6.	Which of the following describes your compressor?
		_____ a. Bad valves (low amps)
		_____ b. Partial short (high amps)
		_____ c. No problem (about right)

CHANGE REFRIGERANT IN A SYSTEM

STUDY MATERIAL
Chapter 6, Unit 4

LABORATORY NOTES

Today's "smorgasbord" of new refrigerants is not only hard to comprehend, but also continuously changing. The safest procedure is to use only original virgin refrigerant in any system. This means that a typical service company will have to obtain and carry many types of refrigerant. Some will carry a certain refrigerant used only in one or a very few systems the company works on.

The next level of refrigerant replacement involves the original equipment manufacturer making a change in the systems they produce to include a new refrigerant and perhaps different oil. The manufacturer will then write up a standard changeover procedure and often make a kit available for doing such jobs as a compressor replacement. The third level of refrigerant replacement involves a field replacement procedure. Manufacturers of the replacement refrigerants are the ones that promote this attitude in the hope of selling refrigerant. This laboratory worksheet should serve to help you determine which level of refrigerant replacement is appropriate in your situation. Since level 1, using only the original equipment manufacturer's original refrigerants and oils is really any normal refrigerant side service, we will divide the jobs into the two categories of original equipment manufacturer approved or field replacement.

OEM APPROVED OR RECOMMENDED REFRIGERANT REPLACEMENT

FIELD REFRIGERANT REPLACEMENT

UNIT DATA

1. Shop ID # _____

2. Unit make _____   Model # _____

3. Unit description _____

4. Original system refrigerant.   Type _____   Amount _____

5. Original oil type (circle one):   Mineral   Alkabenz   Ester

6. Replacement oil type (circle one):   Mineral   Alkabenz   Ester

7. Note all changes to be made besides the refrigerant (circle all that apply):

    Oil   Compressor   Other _____

    _____

8. Describe basic system type.

9. Application and temperature range. _____

_____

10. Obtain job sheet for job to be performed. List and attach sheet.

    Comments. _____

    _____

    _____

ICE MACHINE CLEAN AND CHECK

STUDY MATERIAL
Chapter 18

LABORATORY NOTES

Commercial ice machines are generally accepted to be high maintenance equipment. In addition to the refrigeration side problems, there is the water or ice making side. Any equipment using water is subject to lime buildup, dirt accumulation, freezing, and general water flow problems. Ice machines depend on a steady supply of clean water flow to make good ice. The term "good ice" is used frequently. But what is good ice? It is clear (not cloudy), uniform in shape, hard, (not soft or mushy), and loose (not all frozen together).

Ice machines are typically installed in a small room off to the side of the kitchen of a restaurant. The small room size causes heat and pressure problems. They are in constant use, all day and all night in hot weather. Ice is a product that is not needed too much in cold weather but the minute warm weather hits, the customer needs all they can get. The equipment must not be allowed to freeze due to the water lines within the machine and the water supply lines.

UNIT DATA

1. Ice machine make _____   Model # _____

2. Condenser type (circle one):   Air   Water

3. Measure twenty-four hour ice capacity in pounds.   Weight = _____ lbs

4. Body type (circle one):   Self contained   Remote

ICE INSPECTION

Check	Step	Procedure
_____	1.	Obtain normal system operation as well as possible.
_____	2.	Inspect ice in bin for uniform quality, clear and hard appearance, normal cube separation, etc. To perform this task you must be familiar with the normal size, thickness, and shape of the desired cube shape. The customer may desire a thicker or thinner cube.
_____	3.	Comment on ice inspection. _____

SYSTEM INSPECTION

Check	Step	Procedure
_____	1.	Inspect refrigeration system components for cleanliness and condition.
_____	2.	Compressor inspection (wires, lines, noise). _____
_____	3.	Condenser and fan inspection (dirt, shroud, fan operation). _____
_____	4.	Evaporator inspection (water flow over evaporator). _____
_____	5.	Metering device inspection (suction line, frost, or superheat). _____

SYSTEM CLEANING

Check	Step	Procedure
_____	1.	Obtain cleaning solution recommended by the manufacturer. For nickel plated evaporators check that the solution is safe for nickel.
_____	2.	Shut machine down and empty bin.
_____	3.	Remove any ice from the evaporator. Run water pump with warm water.
_____	4.	Obtain water pump only operation. There is usually a clean and ice toggle switch.
_____	5.	Fill the system with an almost normal water amount.
_____	6.	Dump in 6–8 oz of cleaner. Determine amount by water volume and acid instructions from manufacturer.
_____	7.	Allow water/acid circulation for 1/2 hr to 1 hr.
_____	8.	Using rubber gloves as desired, a sponge, a scraper, clean cloths, diluted acid and soap, clean the bin and ice storage area of any dirt or lime deposit.
_____	9.	Flush water through system three or four times with fresh water.

SYSTEM STARTUP

Check	Step	Procedure
_____	1.	Install gauges and record the system idle pressures.
		High side pressure = _____ Low side pressure = _____
_____	2.	Turn system on, record time. Time = _____
_____	3.	Observe both compressor operation and water pump operation.
_____	4.	Pressures will be low until full water flow is established and then be higher than normal.

Check	Step	Procedure

_____ 5. Record system pressures when the first ice is noticed on the evaporator.

High side pressure = _____   Low side pressure = _____

_____ 6. Measure suction line temperature and liquid line temperature at this time.

Suction line temperature = _____   Liquid line temperature = _____

_____ 7. Calculate system operating superheat and subcooling.

Superheat = _____   Subcooling = _____

_____ 8. Make a judgment on charge based on high side pressure, low side pressure, suction superheat, and subcooling. Which of the following classifies the state of charge?

(circle one):   Overcharged   Undercharged   Correctly charged

Why?_____

_____

_____ 9. Record the time when the system calls for defrost or harvest.   Time = _____

_____ 10. Record the time the harvest is complete.   Time = _____

_____ 11. Allow system operation for several freeze and harvest cycles.

_____ 12. Measure and record average freeze time and harvest time.

Freeze = _____   Harvest = _____

_____ 13. Comment on ice quality and record average freeze time and harvest time. _____

_____

GAS HEATING BOILER LABORATORIES

GAS BOILER STARTUP

STUDY MATERIAL
Chapter 19, Unit 3

LABORATORY NOTES

The gas boiler startup, similar to any other startup, is a turn on, check, and observe job without doing any actual repairs. A heating boiler is a boiler used to heat water for building heat. The temperature can vary from application to application; 180°F for typical baseboard heat, as high as 220°F for certain heating coils, and as low at 120°F for radiant systems. The most common residential system is a single pump multiple zone valve system while commercial systems usually use a separate pump for each zone. All modern hot water boilers have pumps, flow control valves, and an expansion tank. Boilers could be trimmed for steam, meaning the boiler piping is designed to distribute steam. Steam boilers use a gauge glass at the boiler water level, a float operated water level control, steam supply valves, and larger condensate return lines.

UNIT DATA

1. Make _____ Model # _____

2. Boiler rated operating pressure = _____ Pressure relief valve (PRV) = _____

   Opening pressure = _____

3. Burner type _____ Number of burner orifices = _____

4. Type of vent system (circle one): Induced draft   Atmospheric   Power

VISUAL INSPECTION OF BOILER

Check	Step	Procedure
_____	1.	Gas line complete, connections tight and leak free.
_____	2.	Vent connector complete with three screws per joint.
_____	3.	Water lines complete and free from leaks.
_____	4.	Electrical connections complete.
_____	5.	How many water circulating pumps and what type does the system have?

Type _____ Number of pumps = _____

Check	Step	Procedure
_____	6.	How many and what brand of zone valves does the system have?
		Make _____ Model # _____
_____	7.	Any drips or puddles of oil or water? (circle one): Yes No
		If Yes, describe location. _____
		_____
_____	8.	Pump motor and motor mounts centered properly within cradle. Look at motor end view.
_____	9.	Is boiler currently operating? (circle one): Yes No
_____	10.	Locate boiler water temperature pressure gauge and read and record water temperature and pressure.
		Temperature = _____ Pressure = _____

BOILER STARTUP AND OPERATION

Check	Step	Procedure
_____	1.	Turn off all thermostats, observe that burner and pumps are off.
_____	2.	Light pilot as required.
_____	3.	Turn one thermostat to a call for heat.
_____	4.	Observe burner on and pump on.
_____	5.	Observe and feel hot/warm water circulating through that zone.
_____	6.	Observe burner shut down as water reaches set point (180°F).
_____	7.	Record water set point and measure actual water temperature.
		Water set point = _____ Water temperature = _____
_____	8.	Turn first zone thermostat off and each other zone thermostat to a call for heat.
_____	9.	Verify that each zone opens, passes water, and starts both pump and burner.
_____	10.	Listen and observe for any unusual sounds or operation of pump, burner, boiler, etc.

GAS BOILER PREVENTIVE MAINTENANCE

STUDY MATERIAL
Chapter 19, Unit 4

LABORATORY NOTES

This lab leads the student through a typical gas heating boiler inspection preventive maintenance procedure. Heating boilers are identified by the tridecator (triple duty pressure, temperature, and water lift gauge), expansion tank, and water pumps. There are a few older gravity and open expansion tank system but they are pretty rare in today's world. Gravity systems resemble steam systems with the larger lines that slope in the direction of water flow. Steam systems will have a water level control in the boiler, a pressure gauge mounted on a pigtail pipe, to control boiler operation, and all lines slope back to the boiler.

There are a lot of variations in system types in piping and operation and this basic worksheet does not begin to cover all of them. It includes tracing out the system, flushing the boiler to get dirt and mud out, bleeding air from a high point bleeder, and flushing or purging air out of a zone not equipped with high point bleeders. It also includes check and adjusting water pressure in the boiler and the water level in the expansion tank. Both conventional expansion tanks and bladder type tanks are used both in residential and light commercial applications and they need to be checked differently.

UNIT DATA

1. Make _____ Model # _____

2. Boiler maximum operating pressure = PSIG   Hot water = _____   Steam = _____

3. Pressure relief valve (PRV) rating pressure = _____ PSIG   Steam = _____   Pipe size = _____

4. Boiler rated maximum input = _____ Btu/hr

5. Divide maximum input in Btu/hr by 35,000 to obtain boiler HP.   Boiler HP = _____

SYSTEM VISUAL DATA INSPECTION

1. Burner type _____   Number of burner orifices = _____

2. Type of vent system (circle one):   Induced draft   Atmospheric   Power

3. Zone valve information.   Make _____   Model # _____

   Number of zones = _____   Size of valves = _____

Check	Step	Procedure

4. Water pump information.   Make _____   Model # _____

   Number of pumps = _____   Size pumps = _____

5. Type of expansion tank. (circle one):   Conventional tank   Bladder tank

BASIC SYSTEM INSPECTION

Check	Step	Procedure
_____	1.	Gas line complete, connections tight and leak free.
_____	2.	Vent connector complete with three screws per joint.
_____	3.	Water lines complete and free from leaks.
_____	4.	Electrical connections complete.
_____	5.	Pump coupling free of excess water or oil.
_____	6.	Pump motor mounts centered properly within cradle.

CHECK WATER LEVEL IN CONVENTIONAL EXPANSION TANK

Check	Step	Procedure
_____	1.	Observe and record normal boiler operating temperature, pressure, and altitude.

   Temperature = _____ °F   Pressure = _____ PSIG   Altitude = _____ ft

_____	2.	Obtain normal or close to normal values as required, that is, temperature = 180°F, pressure = 15–25 PSIG.
_____	3.	Inspect expansion tank line for slope up to tank.
_____	4.	Locate and inspect B and G airtrol tank fitting (ATF).
_____	5.	Crack bottom bleed nut on tank fitting. What comes out of fitting?

   (circle one):   Air   Water

| _____ | 6. | If air, bleed air slowly until water comes out. There is now the correct level of water in tank. |
| _____ | 7. | If water comes out of tank, advance to the next procedure. |

ADD AIR TO CONVENTIONAL EXPANSION TANK

Check	Step	Procedure
_____	1.	Turn burner off, turn pumps off.
_____	2.	Locate and turn off valve leading to expansion tank.
_____	3.	Install hose on tank drain hose bib and drain water from tank until water stops flowing.
_____	4.	Open bleeder fitting to let air into tank. The bleeder fitting is the same fitting as water came out of in step 5 of the Check Water Level in Conventional Expansion Tank section. The reason we can put air in through the bleeder fitting is because now that the pressure in the tank is off, air will now go in.

Check	Step	Procedure
_____	5.	Observe more water leave tank as air goes in.
_____	6.	Drain water and add air until tank is nearly empty.
_____	7.	Close tank drain valve.
_____	8.	Open tank line valve slowly and feel boiler water enter tank.
_____	9.	Observe boiler pressure drop as water enters tank.
_____	10.	Observe boiler pressure gauge and put your hand on the boiler water feed valve as pressure drops.
_____	11.	Record pressure at which water enters the boiler system. Pressure = _____ PSIG
_____	12.	Hear/feel air leaving tank at bleed fitting.
_____	13.	When water begins to come out of bleed fitting, the water is at the correct level in tank.
_____	14.	Repeat the procedure as required to obtain the correct water level in the expansion tank.

CHECK AIR PRESSURE IN BLADDER TYPE EXPANSION TANK

Check	Step	Procedure
_____	1.	Turn valve off at the tank or bleed down boiler water to 0 PSIG.
_____	2.	Remove tank and drain all water from tank.
_____	3.	If water stays in tank the bladder is leaking and the tank needs to be replaced.
_____	4.	Measure air pressure in tank. Obtain 12 PSIG in tank.
_____	5.	Install cap on tank and check for leaks.
_____	6.	Reinstall tank into boiler system.

ADJUST WATER FEED PRESSURE TO 12 PSIG (IF REQUIRED)

Check	Step	Procedure
_____	1.	Valve off expansion tank and turn burner off.
_____	2.	Connect hose to boiler drain fitting.
_____	3.	Drain water from boiler until fresh water enters. Record the pressure at the water entering point. Pressure = _____ PSIG
_____	4.	Loosen locknut on adjustment screw at top of pressure relief valve (PRV).
_____	5.	Turn adjustment stem in to increase pressure.
_____	6.	Turn water feed pressure off to allow boiler pressure to drop as water continues to drain.
_____	7.	Observe water enter boiler system until the pressure reaches 12 PSIG.
_____	8.	Repeat as required to get a 12 PSIG minimum pressure.

BLEED OR PURGE AIR FROM LINES AND ZONES

Check	Step	Procedure
_____	1.	Check that the system is equipped with high point vent valves. Prepare for the high point bleed.
_____	2.	Turn system off.
_____	3.	Allow 5 min for air to settle out in high points.
_____	4.	Locate all air bleed valves, typically installed on an elbow (coin vent tee) at one end of a section of baseboard or radiator.
_____	5.	Crack each valve open and bleed air until water comes out.
_____	6.	Manually open make-up bypass to get system water pressure to 25 PSIG and repeat step 3.
_____	7.	Turn system on and run.
_____	8.	Shut system down and repeat bleed process until all water is gone and noise from water is gone.

HOSE BIB PURGE SYSTEM

This procedure is for hose bib purge systems, those that are equipped with an isolation valve and a hose bib on each zone.

Check	Step	Procedure
_____	1.	Turn system off.
_____	2.	Turn off all zone isolation valves except one.
_____	3.	Connect hose to hose bib at isolation valve.
_____	4.	Open hose bib and run water into drain.
_____	5.	Hear/notice bubbles of air coming through hose.
_____	6.	Run water through zone until water comes clean.
_____	7.	Close that hose bib and isolation valve.
_____	8.	Move hose to next zone and repeat process until all zones are clear of air.

BURNER SERVICE

Check	Step	Procedure
_____	1.	Turn manual fuel line off.
_____	2.	Remove main and pilot burners. Loosen gas manifold brackets as required.
_____	3.	Keep order of main burners in sequence in larger burners.
_____	4.	Use hacksaw blade, drill orifice, or whatever tool is appropriate to clean rust or debris from burner outlet.
_____	5.	Blow out inside of burners with air pressure or CO_2.
_____	6.	Blow out pilot inside and out.

Check	Step	Procedure
_____	7.	Use a small wire brush to clean debris from burners.
_____	8.	Remove and clean flame rod with steel wool.
_____	9.	Replace any questionable thermocouples.
_____	10.	Be very careful with any glow coil. These parts crack very easily and will likely need to be replaced after a complete burner preventive maintenance.
_____	11.	Refer to original manufacturer literature for positioning of electric spark pilot assembly components, shield, spark probe, and flame sensor.
_____	12.	Reassemble burner components in original positioning.

BOILER STARTUP

Check	Step	Procedure
_____	1.	Boiler is now filled with water and air removed.
_____	2.	Makeup water valve is open.
_____	3.	Boiler is at 12 PSIG idle and cool.
_____	4.	Burner is assembled and ready to be fired.
_____	5.	Expansion tank is ready and connected to system.
_____	6.	Turn power and main fuel on.
_____	7.	Light pilot as required.
_____	8.	Turn thermostat to a call for heat.
_____	9.	Observe main burner on and pump operation.
_____	10.	Record boiler water temperature and observe temperature rising. Pressure = _____ Temperature = _____
_____	11.	Turn on each thermostat separately and observe burner and pump operation during call for heat.
_____	12.	Record final water operating condition. Pressure = _____ Temperature = _____

PREPARE BOILER FOR INSPECTION

STUDY MATERIAL
Chapter 19, Unit 4

LABORATORY NOTES

Occasionally a boiler inspector will request to inspect boiler. The inspector may be a state inspector or an insurance inspector. To inspect a boiler the boiler inspector needs to see the inside of the heating surface of the boiler, and the inside of all safety and operating controls. This job occurs more for steam boilers than for hot water boilers but the procedure is the same. Shut it down, let it cool, pull all caps and plugs, expose the inside of the boiler and the float chamber or steam side mechanism of any moving part. Generally all of these parts are cleaned before inspection.

Boilers tend to accumulate mud, sludge and lime deposits on any inner surface during operation. The inspection is to see how much of this stuff is there. If there is any significant amount of lime or sludge, the inspector will request that it be removed. He must either wait to inspect again or come back later. Either way the boiler may not be put back into operation until it passes inspection. To prepare a boiler for inspection means to clean all or most mud, sludge, lime and any deposits from the inner working of the boiler and its controls. The goal is for the inspector to arrive, pull out a flashlight, look inside, copy down the boiler nameplate data needed and say, "looks good, put it back together."

UNIT DATA AND SYSTEM INSPECTION

1. Make _____   Model # _____

2. Boiler maximum operating pressure = _____

3. Pressure relief valve (PRV) relief pressure = _____

4. Boiler rated maximum input.   Btu/hr = _____

5. Divide boiler input in Btu/hr by 35,000 to obtain boiler HP.   Boiler HP = _____

6. Burner type _____   Number of burner orifices = _____

7. Heat exchanger construction material (circle one):   Steel   Cast iron   Copper   Other____

8. Type of vent system (circle one):   Induced draft   Atmospheric   Power

9. Type of expansion tank (circle one):   Conventional tank   Bladder tank

VISUAL INSPECTION OF BOILER

Check	Step	Procedure
_____	1.	Gas line complete, connections tight and leak free.
_____	2.	Vent connector complete with three screws per joint.
_____	3.	Water lines complete and free from leaks.
_____	4.	Electrical connections complete.
_____	5.	Is boiler is operating? (circle one): Yes No
_____	6.	Record current water values. Temperature = _____ Pressure = _____

BOILER SHUTDOWN AND CLEANING

Check	Step	Procedure
_____	1.	Turn off, lock out, and tag main electrical supply.
_____	2.	Turn off manual main fuel valve.
_____	3.	Turn off boiler water supply.
_____	4.	Allow sufficient cooldown time if boiler is hot.
_____	5.	Turn off all zone line supply valves.
_____	6.	Attach drain down hose as required.
_____	7.	Open boiler drain valve.
_____	8.	Observe water leave boiler.
_____	9.	Open main water supply. Add fresh water. Drain again.
_____	10.	Repeat as required to get clean water from boiler.
_____	11.	Pull bottom or front assembly off low water cut-off and any other boiler control exposed to water or steam.
_____	12.	Remove one cap or plug from below water line (NOWL).
_____	13.	Perform your own inspection of what the inspector will see, use a scraper or spoon to take a sample of any contamination found.
_____	14.	Obtain inhibited acid or other boiler treatment compound, install in boiler as per manufacturer instructions.
_____	15.	Drain, flush, and reinspect all inner surfaces.
_____	16.	Remove all accessible caps and plugs to expose inner surfaces for actual inspection.
_____	17.	Record and perform any additional cleaning, control changes, water treatment, etc., as required by inspector.

PLACE BOILER BACK INTO SERVICE

Check	Step	Procedure
_____	1.	Reassemble all parts in reverse order, replace any required gaskets as required.
_____	2.	Fill boiler to 12 PSIG, bleed air from all lines, baseboard, radiators, and other terminal units.
_____	3.	Remove lockout tag.
_____	4.	Open main fuel valve.
_____	5.	Perform a mini-startup, following the procedure in Laboratory Worksheet GB-1, to verify normal boiler operation.

PUMP SERVICES

STUDY MATERIAL
Chapter 19, Unit 4

LABORATORY NOTES

Hot water heating boilers are dependant on water pumps to distribute the hot water they produce. The pumps are primarily driven by electric motors but occasionally are driven by gas, steam engines. The electric motor requires typical motor service, cleaning, lubrication, proper mounting, checking amperage, etc. The pump is usually connected to the pump with a coupler of either a spider type coupler or a flexible spring connected coupler. Improper alignment is the biggest reason for any coupler problems. Actual pump service required can be generally divided into two categories; impeller problems and shaft seal problems.

On smaller pumps it is typical to exchange the entire pump or bearing assembly for a new or rebuilt replacement. Larger pumps can generally be brought in to be rebuilt by the local factory representatives. An appointment can be made to speed up the rebuild process and get the pump back on line more quickly. To actually rebuild the pump or bearing assembly on site is beyond the scope of this laboratory worksheet. To be able to rebuild on site it would be recommended that the technician, attend a factory service demonstration, work with someone who has done one before, obtain the complete factory rebuild package, and perform rebuilds often enough to stay familiar with the process. Most service technicians don't do this often enough to be experts.

PUMP AND BEARING ASSEMBLY DATA

1. Pump Make _____   Model # _____

2. Motor make _____   HP _____   Type and size of coupler _____

INSPECTION AND CLEANING OF PUMP AND MOTOR MOUNTS

Check	Step	Procedure
_____	1.	Use CO_2 or air pressure blow gun to clean all open air passages of motor and bearing assembly.
_____	2.	Use #20 nondetergent or recommended lubrication oil all motor and bearing assembly bearing as required.
_____	3.	Using a pump sprayer, apply a cleaner degreaser to motor and pump assembly.
_____	4.	Use cleaning rags to wipe off all excess oil and any accumulated dirt, dust, grease, etc.

Check	Step	Procedure
_____	5.	Inspect bearing assembly for any water dripping, metal shavings, scraping or grinding noises, etc. If any, describe. _____
_____	6.	Inspect back motor mounts. Is motor centered within rubber mount or is motor sagging in rubber? _____
_____	7.	If motor is sagging or out of center in the rubber mount, the motor mounts must be replaced.

REPLACE MOTOR MOUNTS

Check	Step	Procedure
_____	1.	Turn off main electrical supply.
_____	2.	Using a thin open end wrench, loosen two top motor mount bolts from inside the bearing assembly.
_____	3.	Loosen and remove the two bottom motor mount bolts.
_____	4.	Use a long or T-handle Allen wrench, to remove the motor end of the spring coupler.
_____	5.	Supporting the motor in one hand, remove the previously loosened two top motor mount bolts.
_____	6.	Remove motor from its cradle, two machine screws connected to motor mount straps.
_____	7.	Using an old screwdriver or a thin chisel pry off both rubber motor mounts from the motor.
_____	8.	Use a small metal, rubber, or plastic hammer to tap new rubber motor mounts.

REPLACE BEARING ASSEMBLY

Check	Step	Procedure
_____	1.	Turn off water isolation valves above and below pump and bearing assembly. If pump does not have isolation valves, reduce boiler pressure to 0 PSIG and turn off every valve you can find.
_____	2.	Remove motor as in step 3 of the Replace Motor Mounts section.
_____	3.	Remove four bolts holding bearing assembly to pump housing.
_____	4.	Remove bearing assembly with pump impeller.
_____	5.	Inspect and clean any debris from pump housing.
_____	6.	Holding impeller with a strap wrench, use a socket wrench to remove bolt holding impeller on pump shaft.
_____	7.	Inspect impeller, replace if any signs of wear or deterioration.
_____	8.	Install impeller on new bearing assembly.
_____	9.	Remove old gasket from bearing assembly. Carefully scrape clean surface metal.
_____	10.	Clean matching metal surface on pump housing.
_____	11.	Install new gasket and slip bearing assembly into pump housing.

_____ 12.   Start all four pump bolts, and then snug with wrench.

_____ 13.   Install spring coupler on bearing assembly shaft.

_____ 14.   Support motor in position and slide spring coupler onto motor shaft. Be sure to set screws into hollow spot on both pump and motor shafts.

_____ 15.   While still supporting motor, start two top motor mount bolts first, then two bottom bolts.

_____ 16.   Snug four motor mount bolts with open end wrench.

_____ 17.   Open all water valves, obtain normal pressure on system, bleed or purge air from system as required.

_____ 18.   Start up and obtain normal operation.

_____ 19.   Lubricate bearing assembly and motor with oil as required.

_____ 20.   Inspect bearing assembly for water leaks.